Scenic Driving

CONNECTICUT AND RHODE ISLAND

Scenic Driving

CONNECTICUT AND RHODE ISLAND

Exploring the States' Most Spectacular Byways and Back Roads

STEWART M. GREEN

Guilford, Connecticut

Globe
Pequot

An imprint of Rowman & Littlefield

Distributed by NATIONAL BOOK NETWORK

Photography by Stewart M. Green

Excerpted from *Scenic Routes & Byways New England* (Globe Pequot, 978-0-7627-7955-0)

British Library Cataloguing in Publication Information Available
Library of Congress Cataloging-in-Publication Data is available on file.

ISBN 978-1-4930-2237-3 (paperback)
ISBN 978-1-4930-2238-0 (e-book)

∞™ The paper used in this publication meets the minimum requirements of American National Standard for Information Sciences—Permanence of Paper for Printed Library Materials, ANSI/NISO Z39.48-1992.

TABLE OF CONTENTS

Connecticut & Rhode Island

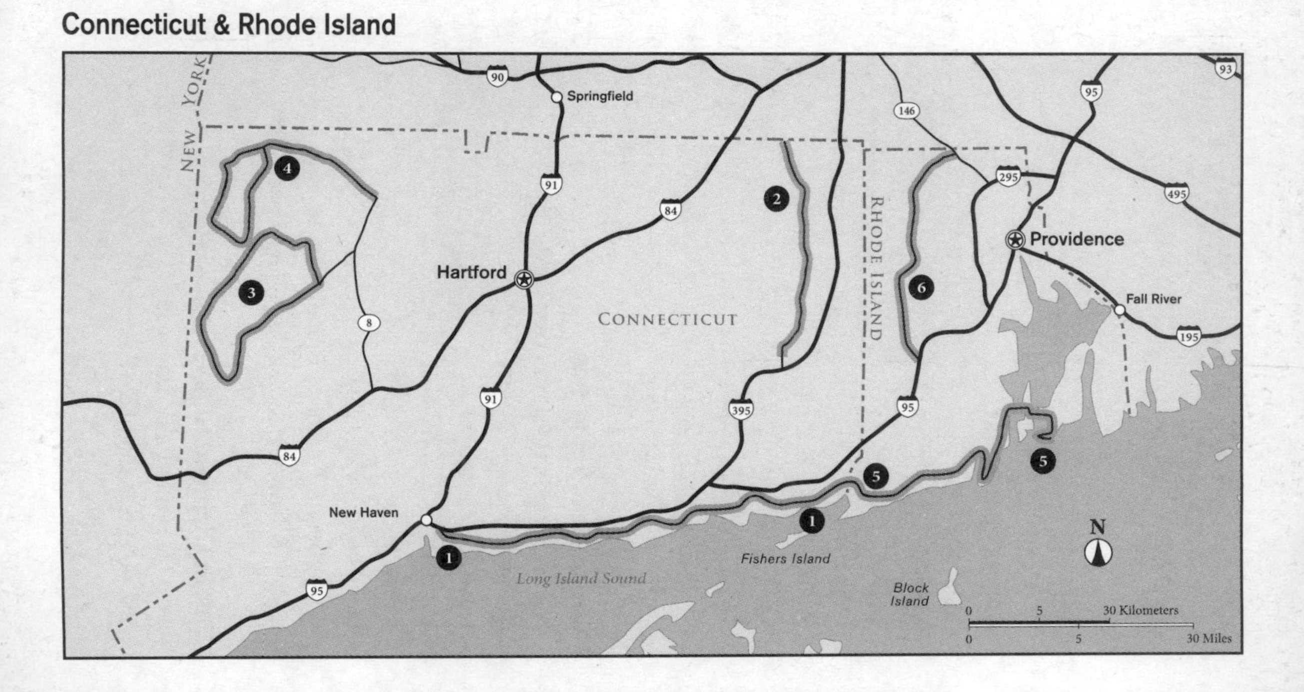

INTRODUCTION

Connecticut and Rhode Island offer travelers a spectacular assortment of natural and scenic wonders, historic sites, and varied recreational opportunities. Numerous state parks, forests, beaches, and recreation areas preserve slices of superlative landscapes. In addition to the ocean, hundreds of lakes and ponds and numerous rivers and brooks offer boating, swimming, canoeing, and angling choices for outdoor enthusiasts. The cities, towns, and villages, from the urban centers of Hartford and Providence, to myriad tiny villages dotting the hills of Connecticut and Rhode Island, are filled with culture and steeped in history.

Scenic Driving Connecticut and Rhode Island, an indispensable mile-by-mile highway companion, explores and discovers the wonders of this compact region. The drives follow miles of highways and back roads, sampling the region's colorful historical sites, beauty spots, hidden wonders, and scenic jewels. Drivers will wind along ragged headlands pounded by the restless ocean along the Connecticut coast, marvel at classic villages set among beaches and hills, pass rural birthplaces and burial sites of the notable and the notorious, and wander among shifting sand dunes. Most of the drives leave the urban sprawl and interstate highways behind, setting off into the beautiful heart of the region.

Connecticut and Rhode Island are laced with highways and roads, some dating back to the earliest paths that once connected colonial settlements. Area natives will undoubtedly wonder why some roads are included and others omitted. These routes were chosen for their beauty, unique natural history, and historical implications. Omitted are worthy roads for one reason or another, but mostly due to the burgeoning development along those asphalt corridors in an amazing labyrinth of highway possibilities.

Use these described drives to win a new appreciation and understanding of this marvelous landscape. Take them as a starting point to embark on new adventures by seeking out other backroad gems among the rolling hills and vales of Connecticut, in the historic towns and old hills of Rhode Island.

Travel Advice

Be prepared for changing weather when traveling these scenic highways, especially in winter when snow and ice encase the roadways. Most of the drives, except for bits and pieces, are paved two-lane highways that are regularly maintained. Services are available on all the drives, and every little village offers at least some basics during daylight hours. Use caution when driving. Many of the roads twist and wind through valleys and over mountains, with blind corners. Follow the posted speed limits and stay in your lane. Use occasional pullouts to allow faster traffic to safely pass. Watch for heavy traffic on some roads, particularly during summer vacation season and on fall-foliage weekends. Be extremely alert for animals crossing the asphalt. Take care at dusk, just after darkness falls, and in the early morning.

The region's fickle weather creates changeable and dangerous driving conditions. Make sure your windshield wipers are in good shape. Heavy rain can impair highway vision and cause your vehicle to hydroplane. Snow and ice slicken mountain highways. Slow down, carry chains and a shovel, and have spare clothes and a sleeping bag when traveling in winter. Watch for fog and poor visibility, particularly along the coastlines. Know your vehicle and its limits when traveling and, above all, use common sense.

Travelers are, unfortunately, potential crime victims. Use caution when driving in urban areas or popular tourist destinations. Keep all valuables, including wallets, purses, cameras, and video cameras, out of sight in a parked car. Better yet, take them with you when leaving the vehicle.

These drives cross a complex mosaic of private and public land. Respect private property rights by not trespassing or crossing fences.

Remember also that all archaeological and historic sites are protected by federal law. Campers should try to use established campgrounds or campsites whenever possible to avoid adverse environmental impacts. Remember to douse your campfires and to pack all your trash out with you to the nearest refuse container.

Every road we travel offers its own promise and special rewards. Remember Walt Whitman's poetic proclamation as you drive along these scenic highways: "Afoot, light-hearted, I take to the open road. Healthy, free, the world before me."

Legend

95	Interstate Highway
44	US Highway
41	State Highway
	Local Road
	Featured Route
	Trail
	State Border
	River/Creek
	Lake
	Capital
	City
	Town
	Bridge
	Campground
Exit	Exit Sign
	Peak
	Point of Interest
	State Park, Indian Reservation, Management Area

1 Connecticut Coast

General description: A 71-mile scenic route along southeastern Connecticut's historic coastline.

Special attractions: Mystic Seaport: The Museum of America and the Sea, Mystic Aquarium, Bluff Point State Park, Fort Griswold Battlefield State Park, US Navy Submarine Force Museum, Rocky Neck State Park, Old Lyme, Lyme Academy of Fine Arts, Old Saybrook, Florence Griswold Museum, Hammonasset Beach State Park, Guilford, Whitfield House, historic houses and sites, picnicking, camping, fishing, swimming, beaches.

Location: Southeastern Connecticut.

Drive route numbers: US 1, I-95, CT 156 and 146.

Travel season: Year-round.

Camping: Rocky Neck State Park has a 160-site campground that is open from mid-April through the end of September. Hammonasset Beach State Park has the William F. Miller Campground with 558 sites and beach access. Call (203) 245-1817 for park campground information and reservations.

Services: All services in Pawcatuck, Mystic, Groton, New London, Niantic, Old Lyme, Old Saybrook, Clinton, Madison, and Guilford.

Nearby attractions: Watch Hill (RI), Misquamicut State Beach (RI), Nehantic State Forest, Becket Hill State Park, Connecticut River, Selden Neck State Park, Gillette Castle State Park, East Haddam, Brainard Homestead State Park, Haddam Meadows State Park, Cockaponset State Forest, Chatfield Hollow State Park, New Haven attractions.

The Route

This scenic route traverses 71 miles of southeastern Connecticut's coastline, passing numerous historic areas and towns, and crossing the unspoiled mouth of New England's longest river while offering tranquil vistas of sandy beaches, salt marshes, and dense

Connecticut Coast

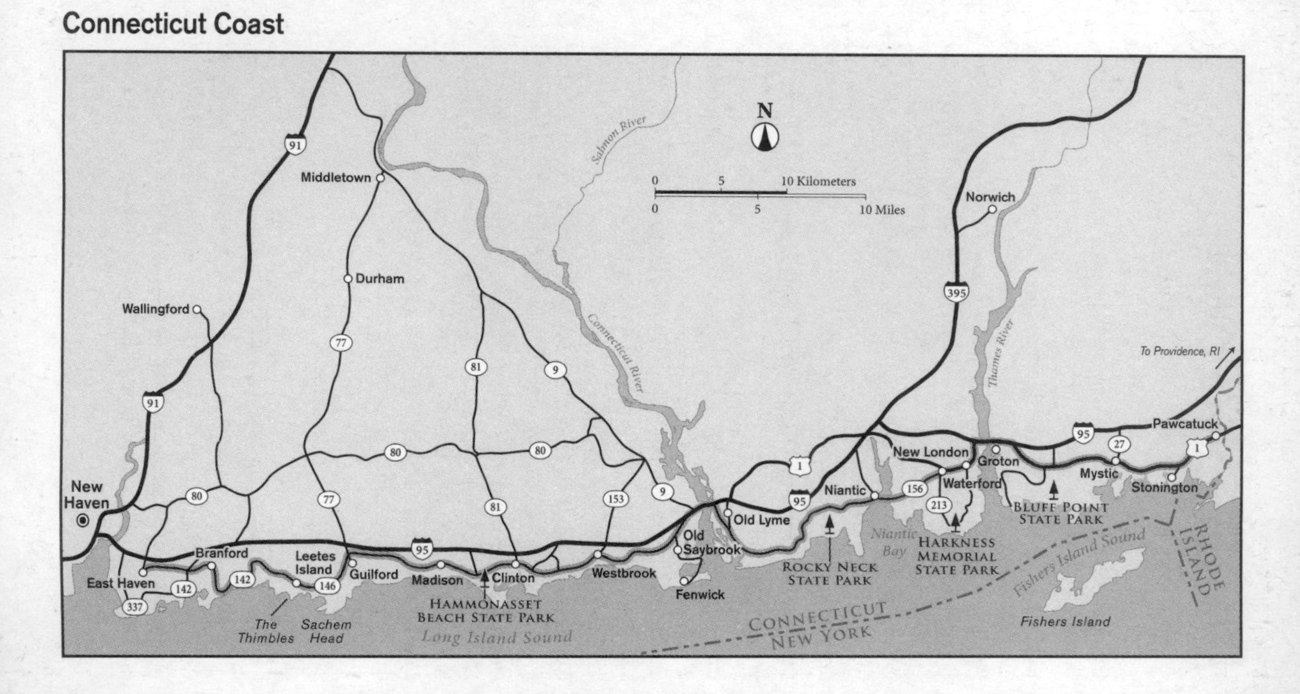

woodlands. Worlds away from the Nutmeg State's busy urban centers, this lovely drive parallels coves, inlets, and beaches along Long Island Sound between Rhode Island and the outskirts of New Haven. This section of coast boasts a long and colorful seafaring history, with many of its towns, including some of the state's oldest villages, drawing sustenance from the ocean.

The 19th-century Industrial Revolution, which transformed and urbanized much of southern New England, mostly bypassed the fishing and whaling villages along the southeastern Connecticut coast. Instead, industry developed inland at the fall line of the region's rivers, with rapids and waterfalls generating power for mills and factories.

Today this shoreline, lying just off I-95 and the hectic Boston–New York corridor, is generally overlooked by passing drivers. Those who do venture off the interstate are treated to quiet country roads and a look at Connecticut's rich maritime heritage.

Heading west from Rhode Island to New Haven, the drive begins in Pawcatuck on the west side of the Pawcatuck River, across from Westerly in Rhode Island. Pawcatuck is easily reached from I-95. Take exit 2 and drive south a few miles on CT 2 to its junction with US 1, the East Coast's celebrated main thoroughfare. US 1 roughly follows the old Boston Post Road, the colonial road between Boston and New York. Head west through Pawcatuck on US 1 past many large homes and leave town after a mile. A few miles later is US 1A; jog left here to Stonington.

Stonington

One of Connecticut's loveliest coastal villages, **Stonington** sits on a narrow peninsula jutting south into Fishers Island Sound. The town, dubbed the "Nursery for Seamen," offers a superb, protected harbor that nourished its profitable whaling, sealing, fishing, trading, and shipbuilding industries through the 19th century. This once-busy town, settled in the 1640s, figured prominently in early

Connecticut history when the local militia repelled British naval attacks during both the Revolution and the War of 1812.

The latter battle is commemorated by a historical tablet on Water Street that reads: "The brave men of Stonington defeated a landing force from the British ship *Ramillies* bent on burning the town and its shipping. August 10, 1814." Cannon Square holds a pair of 18-pound cannons that the patriots used to repulse four British warships, which were commanded by Captain Thomas M. Hardy and armed with 140 guns. The invaders sent a note ashore warning the villagers to abandon the town within one hour or face their wrath. Stonington's soldiers replied, "We shall defend the place to the last extremity; should it be destroyed, we shall perish in its ruins." Captain Jeremiah Holmes tacked an American flag over the town's earthen breastworks, averring, "That flag shall never come down while I'm alive." The subsequent 2-day British barrage injured only one person, while the local cannons killed 21 British sailors, wounded 50 more, and almost sank the *Dispatch* before the ships withdrew.

This seaport town, at one time the third largest in the state, was a bustling commercial hub at its zenith. Stonington's two main avenues were jammed with seamen, whalers, and sailors and lined with buildings, warehouses, and the opulent houses of wealthy merchants and sea captains. Ships based here fanned out around the globe, including the *Betsy,* the first American ship to circumnavigate the world. Captain Nathaniel Parker sailed from here on a sealing expedition in 1820 and discovered Antarctica. The town, with as many as 17 separate tracks, was also an early railroad center with passengers arriving by rail from Boston and departing for New York on ferries.

Stonington still retains its colonial charm with cobblestone streets and many Federal and Greek Revival–style homes. A walking tour brochure details important sites, including homes once

The historic 1852 Captain Nathaniel B. Palmer House preserves a slice of Stonington's seafaring history.

A family enjoys DuBois Beach at the end of Water Street in Stonington.

occupied by poet Stephen Vincent Benét and painter James Whistler. With its narrow streets hemmed in by restored buildings and houses and numerous shops, Stonington is a great place for walkers, rather than drivers, to roam and explore.

Be sure to stop at the **Old Lighthouse Museum** at the end of Water Street, overlooking the Atlantic. The sturdy granite lighthouse, built in 1823, was the first erected by the federal government. Inside is an eclectic collection of local artifacts and memorabilia, including distinctive Stonington salt-glaze pottery, ship models, whaling gear, cannonballs from the Battle of Stonington, and treasures brought back from the Orient. Circular stone stairs lead to a small chamber atop the lighthouse and marvelous views of the surrounding harbor, Block Island to the southeast, and even the tip of Montauk Point on Long Island to the south.

Return to US 1 by taking Water Street north and following signs for US 1 and I-95. Back on US 1, head west past the upper end of placid Stonington Harbor, past a turnoff to Lords Point, and run through a forest dotted with homes. After almost 5 miles

The Stars and Stripes keeps a patriotic front gate closed in old Stonington.

the road bends northwest and enters Mystic, the best known of New England's living history museum-villages.

Mystic

Mystic, named from the Indian word *mistick* or "tidal river," was, along with its neighbor Stonington, one of Connecticut's most prosperous and important seaports. The fabled town, straddling the banks of the **Mystic River** where it empties into Mystic Harbor, flourished not only as a port but as one of the nation's premier shipbuilding centers in the 19th century. Reaching 6 miles inland from the ocean, the wide, calm river channel with its gently sloped banks was ideal for constructing the variety of ships needed by the country's growing clipper trade. The nearby forests of cedar, white pine, ash, spruce, and white oak offered plentiful raw materials. Hundreds of clippers, packets, sloops, and whalers, as well as 56 steamers for the federal navy, were launched from Mystic's shipyards. Mystic's ships often made the hazardous

110-day journey around Tierra del Fuego to San Francisco during the Gold Rush days, and competing shipbuilders vied to build the fastest vessel. The *Andrew Jackson,* built in Mystic in 1860, still holds the record—89 days and 4 hours. The town also flourished as a whaling port from 1832 to 1860 when petroleum began replacing whale oil for illumination and lubrication.

This colorful maritime history is commemorated at **Mystic Seaport: The Museum of America and the Sea** on the waterfront just north of the drive on CT 27. The 17-acre museum re-creates a typical 19th-century New England seaport with more than 60 historic buildings and homes relocated here. On the site of two old shipyards, it is a living history site displaying restored sailing ships that nudge against piers and interpreters who practice traditional maritime crafts. The replica village gives a uniquely vivid glimpse back to the 1860s during the heyday of the tall ships. Visitors ramble through the streets past the town tavern, shops, schoolhouse, church, bank, and print shop, and along the wharves where the ships are anchored. Dozens of exhibits include whaling tools and a video; a collection of figureheads; the Preservation Shipyard where visitors watch a ship restoration; and the N. G. Fish Ship Chandlery, which outfitted sailors and ships for voyages that lasted as long as 5 years.

The ***Charles W. Morgan,*** a 113-foot-long whaleship built in 1841, is the centerpiece of the Mystic fleet. The ship, a national historic landmark, made 37 voyages in 80 years under the command of 21 different captains—and made almost $1.5 million for its New Bedford owners. Other tall ships parked here are the two-masted ***L. A. Dunton*** fishing schooner, the **SS *Sabina*** steamboat, and the ***Joseph Conrad,*** a square-rigged Danish training ship. Mystic Seaport is open year-round. Allow a full day to visit the museum and discover New England's rich maritime heritage.

Pleasure boats rest at anchor in the fabled seaport harbor of Mystic.

Visitors relax at the entrance to Mystic Seaport: The Museum of America and the Sea.

Before leaving Mystic, park and walk around the town. Its sidewalks, often crowded with museum visitors, are lined with shops, cafes, and historic colonial homes. A couple points of interest include the 1764 **Whitehall Mansion,** the 1717 **Denison Homestead** (furnished by 11 generations of Denisons).

The **Mystic Aquarium,** on the north side of town, displays more than 6,000 sea creatures, as well as seals and sea lions. The drive threads through Mystic on US 1 as it crosses the river on a drawbridge, passes through downtown Mystic and West Mystic, and twists west through wooded hills. A few miles later is the turnoff to **Bluff Point State Park.** This 806-acre parkland, encompassing a forested neck that reaches down to Long Island Sound, is the largest remaining stretch of wild, undeveloped coast along Connecticut's 253-mile shoreline. Overlying metamorphic gneiss and schist rocks, the park is a wintering area for waterfowl, including brant and scaup. Nearby is Bluff Point Beach and a narrow sandspit or tombolo that ends at Bushy Point, a small island at high tide.

Groton

Past the park turnoff, the highway enters Groton, one of the larger cities along this stretch of coast. **Groton,** spread along the east bank of the Thames River, was originally part of the Pequot Plantation founded in 1646 by John Winthrop Jr. In 1705 it was incorporated and given the name Groton to honor Winthrop's English estate.

During the American Revolution, Groton was the site of the only major battle in Connecticut, fought at Fort Griswold. This dark episode, one of the most savage and brutal fights of the war, occurred on September 8, 1781. The turncoat traitor Benedict Arnold led a British assault on Groton and New London. When Groton was attacked by two regiments, Lieutenant Colonel William Ledyard mustered 165 militiamen to defend Fort Griswold, a 12-foot-high stone fort overlooking the Thames River. The fort was surrounded, but Griswold's defenders refused to surrender. The British force of 800 began a determined attack but was repulsed several times before breaking into the fort. The British suffered high causalities, 50 killed and more than 150 wounded, while only 3 patriots were killed. Ledyard surrendered his garrison, handing his sword to the opposing commander in subjugation. The Tory officer, irate after the brutal battle, promptly ran Ledyard through with the sword. A slaughter ensued, with the British soldiers massacring 88 of the surrendered Americans after their weapons were laid down.

The story of this infamous encounter is now preserved at **Fort Griswold Battlefield State Park** on Fort Street in Groton, and the memory of the murdered defenders is kept alive with a 134-foot granite obelisk. The names of the slain men are engraved on the monument and a spiral staircase winds to a panoramic viewpoint atop the tower. The star-shaped fort remains in remarkably good shape. An on-site museum offers a diorama of the battle and exhibits of Groton history.

Today Groton is famous as the "submarine capital of the world." The town, with the nation's largest submarine base, is also the birthplace of the nuclear-powered submarine. After building 74 subs during World War II, General Dynamics also built the first nuclear sub, the **USS *Nautilus,*** in 1954. The *Nautilus,* now decommissioned after 25 years of service and almost 500,000 miles, is open for free tours year-round. Visitors can explore this fascinating National Historic Landmark, viewing bunks for the 111 crew members, officers' quarters, the torpedo room, and the attack center. Nearby is the **US Navy Submarine Force Museum,** exhibiting the history of underwater warfare, multimedia shows, and numerous midget subs.

New London

In Groton, follow signs to I-95 South. Get on the interstate, cross the Thames River, and take exit 84 to US 1. The drive and US 1 head south and west through a dense business district where **New London,** one of Connecticut's venerable coastal cities, spreads along the west bank of the Thames River opposite Groton. The town was originally settled in 1646 by a group of 40 Puritan families from Massachusetts led by John Winthrop Jr. It was named for London, England, of course, and in keeping with that theme, the newcomers appropriately renamed the Monhegan River, the Thames River. New London boasted one of the finest deepwater ports on the southern New England seaboard and quickly prospered with West Indies trade in sugar, molasses, and rum.

During the Revolutionary War, New London was the base for some of the war's most notorious privateers—pirates who worked for the government by seizing and capturing enemy ships. The New London privateers raised British ire by their constant harassment. In 1781 Benedict Arnold led a British armada of 32 ships and a military force of 1,700 soldiers against New London and Groton, leading not only to the Fort Griswold massacre but also to the burning and destruction of much of New London. After the war many New London residents migrated west to Ohio and land given by the federal government to compensate for their wartime losses.

The town, along with New Bedford, became a leading whaling port between 1784 and 1909. The whale wealth built **Whale Oil Row,** a line of four Greek Revival mansions built in 1830. Other historic sites in New London include the 78-acre **Downtown New London Historic District** with 223 historic buildings along the waterfront and the granite block **US Customs House,** built in 1833, one of the nation's oldest operating custom houses. The city offers a host of historic buildings and homes, many open for tours, including the 1678 and 1759 **Hempsted Houses** and the 1756 **Shaw Mansion.** One of the most interesting is **Monte Cristo Cottage,** the childhood home of famed American playwright Eugene O'Neill. O'Neill, who won a Nobel Prize and four Pulitzer Prizes, based some of his works on childhood experiences at the house. The **O'Neill Theater Center** runs a museum at the house and mounts productions of his plays every summer.

New London is also home to the **US Coast Guard Academy** on a 125-acre site along the Thames River, the elegant **Lyman Allyn Art Museum** with a fine collection of Connecticut painters, and the informative DNA EpiCenter Inc. with natural history displays of the Thames River basin.

Waterford to Old Lyme

Drive west through New London, following US 1 through a residential district before leaving town. The next 15-mile drive stretch runs from here to I-95 exit 70 at Old Lyme. In just a few miles more, the highway enters **Waterford,** a quiet town that was part of New London from its 1645 settlement until its incorporation as a separate entity in 1801. A collection of historic buildings gather about the **Jordan Village Historic District,** including the Beebe Phillips Farmhouse with exhibits on farm living in the 1800s; the quaint 1749 Jordan School House, Waterford's oldest public edifice; and the Stacy Barn with farm implements and wagons. South of town on Great Neck Road is 230-acre **Harkness Memorial State Park,** which includes a 42-room Italianate villa with formal gardens overlooking Long Island Sound.

Bear southwest (left) in Waterford on CT 156 W to the historic district, the turn to the state park, and to continue the drive. CT 156 leaves the city behind and heads through wooded hills studded with homes. The road skirts Niantic Bay before crossing the Niantic Bay Bridge, a drawbridge above a marina and harbor with a stunning view of the bay. The route then runs through Niantic itself. This peaceful coastal village offers several historic homes, pleasant seaside views, and nearby beaches. The **Thomas Lee House,** open Wednesday to Saturday for spring and fall tours, was built in 1660 and is the oldest remaining wood frame house left in Connecticut. It was originally a one-room cabin that may have been built as early as 1641 before later expansion. The house is furnished with rare 17th-century English furnishings, as well as an original casement window.

After a mile the highway exits Niantic. Look down the road for a couple of miles until you see the left turn to **Rocky Neck State Park.** This popular 710-acre state park offers a dazzling mile-long, white-sand beach, one of a few state-run beaches in Connecticut. The gently sloped beach, warm seawater, and easy interstate access make this a busy summer destination. Off-season it is less crowded.

Four Mile River, a tidal river, borders the park on the west. The river's east side is a large salt marsh that offers great birding opportunities, with common sightings of heron, teal, mallard, and osprey. The park also offers a 160-site campground, 4.5 miles of hiking trails, saltwater fishing, picnic tables, and other recreational facilities.

The drive continues through a thick mixed hardwood forest and after a few miles bends inland along the east bank of the broad Connecticut River to Old Lyme with its marshes and dense thickets. The **Connecticut River,** emptying here into Long Island Sound, is one of the only major East Coast rivers with an undeveloped mouth. The longest river in New England, it flows 407 miles from its source at the Canadian border to the sound, dropping 1,618 feet and draining more than 11,000 square miles. The river's name (and the state's) derives from the Algonquin Indian word Quinatucquet, meaning "on the long tidal river." In the 1600s, the river formed the dividing line between the Dutch colonies along the Hudson River and the English settlements in Massachusetts and Rhode Island.

Old Lyme

Set on the east bank of the Connecticut River, **Old Lyme** is a classic New England village. Its First Congregational Church, built in 1817 and rebuilt and fire-proofed after a 1907 fire, is a beautiful building fronted by Ionic columns and topped with a tall, graceful steeple. The town, which split from Saybrook in 1665, was an old shipbuilding center and home port to many sea captains. It was named for the English port of Lyme Regis.

At the end of the 19th century, however, Old Lyme became a flourishing artists' colony and center of American Impressionism. The gorgeous scenery and bucolic woodlands along the river valley and seacoast attracted many talented painters, including Childe Hassam, Henry Ward Ranger, and William Metcalf, who found the surrounding countryside, light, and color palette similar to that around Paris. One of the Impressionists' favorite subjects was the village church. The **Lyme Academy of Fine Arts,** with exhibits by today's local artists, is nearby. Florence Griswold, spinster daughter of Captain Robert Griswold, opened her inherited Georgian mansion to the landscape painters and became a patron of the arts. The artists adorned her dining room with romantic landscape murals. Today the rest of the 1817 house is an art museum with a collection of works by more than 130 American artists, including many lovely Impressionistic pieces.

Old Saybrook to Guilford

Continue the drive by heading up CT 156 to I-95. Take the interstate across the half-mile-wide Connecticut River before leaving it again at exit 68 to US 1 and Old Saybrook. The next 20-mile drive section runs from here to Guilford.

Old Saybrook has the honor of being the first colonial settlement on New England's southern shore. The village, located at a place the Indians called Pashbeshauke or "place at the river's mouth," was founded in 1635 as Saybrook Colony by Puritan settlers. Old Saybrook was the initial home of Yale University, founded here in 1701 as the Collegiate School before moving to New Haven 15 years later. It was also the home to inventor David Bushnell, who, after putting himself through Yale at the age of 31, built a sea craft he called a "sub-marine" for use in the Revolution. His submersible, dubbed the American Turtle, was meant to maneuver underwater to the hull of enemy ships, upon which would be attached an exploding, timed mine of his invention.

Old Saybrook spreads along Main Street southeast of US 1 to Saybrook Point, a peninsula jutting into the Connecticut River's mouth. A walking tour on Main Street gives a taste of the village's colorful history. The Humphrey Pratt Tavern, on the National Register of Historic Places, was an old stage stop as well as Saybrook's first post office. The 1840 Greek Revival–style Congregational Church is the congregation's fourth building since 1646. The **General William Hart House,** a handsome house-museum open for tours, is a superb example of a pre-Revolution colonial home. The restored house, built in 1767, is surrounded by colonial gardens with fruit trees, roses, and herbs. At Saybrook Point are numerous lovely old homes overlooking the river mouth, along with Cypress Cemetery and the earthen 1636 Saybrook Fort, the state's first fortification.

The drive continues west along US 1, passing through Westbrook, a village settled in 1648. The town's picturesque Congregational Church overlooks the Westbrook Cemetery, with many old

A stony beach at Hammonasset Beach State Park edges Long Island Sound.

gravestones from the 1700s. Nearby is the restored 1756 house of settler Jedediah Chapman. The highway meanders above curving Westbrook Harbor before bending inland to Clinton. This quiet community with tree-lined streets dates from 1663. The 1850 **Captain Elisha White House,** a brick house open on summer weekends, was lovingly restored after a fire gutted it at the turn of the 20th century.

The highway leaves Clinton and crosses the Hammonasset River. The turnoff to **Hammonasset Beach State Park** is just west of the bridge. This 919-acre parkland makes a great stop, offering a 2-mile strand of sandy beach and the state's largest park campground. Like Rocky Neck, it is easily accessible to I-95 and much of Connecticut's population, so it is crowded in summer. The park spreads across a wide neck that reaches into Long Island Sound. A good hike lies along the cobbled shoreline at Meigs Point, with spectacular views south across the sound to the forest-fringed outline of Long Island. The nearby **Meigs Point Nature Center** interprets the area's natural history with nature programs and walks. The extensive marshlands along the river attract shorebirds, waterfowl, and birders off-season.

Madison is the next coastal town on US 1. The town was originally part of Guilford but separated in 1826 and was named for President James Madison. Its stately village green is surrounded by many homes that date from the 1700s along with the ubiquitous Congregational Church topped with a small gold-domed steeple. The house to visit here is the **Allis-Bushnell House** and museum. The two-family house, originally built in 1785, was once home to Cornelius Scranton Bushnell, an organizer of the Union Pacific Railroad, shipbuilding magnate, and financier of the *Monitor,* a famed ironclad ship from the Civil War.

Guilford

Next the highway leaves Madison, passes an old cemetery, and rolls through wooded countryside interrupted by occasional stone walls around pastures with grazing sheep and cattle. Five miles later the drive enters the historic settlement of **Guilford,** the seventh-oldest town in Connecticut. This beautiful village, dating from 1639, surrounds a tree-shaded green lined with old houses, four churches, and the town hall. The green, like all New England village greens, was used as a pioneer burial ground and a place to pasture cattle and horses, flog common criminals, and drill the local militia.

Reverend Henry Whitfield and 40 Puritan men and their families bought land from the Menuncatuck Indians and settled here as farmers. In the 18th century, Guilford prospered through its shipyard, fishing, and trading. During the Revolution it avoided destruction by the British and kept intact its heritage of several 17th-century houses, as well as more than 400 18th- and 19th-century homes.

The 1639 Whitfield House in Guilford is reputed to be the oldest stone house in New England.

The **Henry Whitfield State Museum,** lying south of the green on Old Whitfield Street, is the town's most famous structure. The house, with parts of it built in 1639 by the town's founder, is reputed to be New England's oldest stone dwelling. The house was meant to be not only home to the Whitfields, but also a fortress to protect the first settlers from Indian attacks. Over the last three centuries, the house has changed dramatically, and Whitfield would scarcely recognize his old home. The house does, however, illustrate daily life in early Connecticut with its 33-foot-long Great Hall, leaded glass windows, whitewashed walls, and rare 17th- and 18th-century artifacts and furnishings. Other local houses in Guilford are typical New England saltboxes including the 1660 **Hyland House** and the 1774 **Thomas Griswold House.** Both are restored and operated as museums. On the south side of town fronting the ocean is **Chaffinch Island Park** with a picnic area, Jacobs Beach, and the town landing and marina.

Guilford to New Haven

The last drive segment begins in Guilford and runs 18 miles through Branford to East Haven and I-95. Find this road section by keeping left on CT 146 or Boston Street at its junction with US 1 on the east side of Guilford. After leaving Guilford, the highway is a designated Connecticut Scenic Highway for the next 12 miles. The narrow asphalt road dips and rolls among low rocky hills blanketed with dense forest and skirts salt marshes, inlets, and coves. Occasional ocean views unfold beyond the trees.

Sachems Head, a stubby peninsula jutting into the water south of the highway, possesses a rich history. It was named after a Mohegan chief, Uncas, who killed a captured Pequot enemy sachem, or chief, here and wedged his severed head in the fork of a tree during the Pequot War in the late 1630s. The Pequots, their name meaning "destroyers of men," were the most fierce of the 16 Connecticut tribes. The Pequots and Mohegans had been allied until Uncas was spurned as the leader of the two tribes.

In retaliation he lined up an uneasy alliance with the colonists against the Pequots, who had raided settlements and killed many residents.

Past Island Bay, the road reaches Leetes Island, the site of a 1777 skirmish where the Guilford militia doused a British raiding attempt. The highway continues west within sight of the coastline and offers views of the Thimble Islands archipelago. This offshore group of 365 islands, named for the abundant thimbleberries found on the wooded isles, forms one of the most scenic portions of the Connecticut coast. Unfortunately, they are all privately held and trespassers are not highly regarded. The notorious pirate Captain Kidd supposedly buried a still unfound treasure on one of the Thimble isles in 1699. Several vessels offer daily sightseeing excursions among the islands from **Stony Creek Dock** at the end of Thimble Island Road.

The highway rolls westward through Pine Orchard and Indian Neck before turning north to **Branford.** This village was, like most other coastal towns, an early settlement. It was purchased from Indians in 1638 and settled the following year. Historic houses, three churches including a lovely Congregational Church topped with a steeple and clock tower, and the town hall surround the old town green.

To end the drive, continue west on US 1 from Branford a few miles to its intersection with I-95 and exit 51. An alternate and longer ending follows CT 142 and 337 down along the coast and up the east edge of New Haven Harbor. This winding road passes through numerous residential suburbs amid low granite hills. CT 337 ends at exit 50 on I-95. The New Haven metropolis awaits the traveler immediately west.

2 Connecticut Route 169

General description: The drive, a National Scenic Byway, follows Connecticut Route 169 (CT 169) for 32 miles through historic villages and pastoral countryside in northeastern Connecticut.

Special attractions: Woodstock, Roseland Cottage, Pomfret, Mashamoquet Brook State Park, Brooklyn Green, General Israel Putnam Statue, Prudence Crandall Museum, Canterbury, Newent, historic villages and houses, national historic districts.

Location: Eastern Connecticut. Drive parallels the Rhode Island border and I-395.

Drive route numbers: CT 169.

Travel season: Year-round.

Camping: 2 campgrounds with 55 sites at Mashamoquet Brook State Park.

Services: All services in Woodstock, Pomfret, Brooklyn, Canterbury, and Lisbon.

Nearby attractions: Mohegan State Forest, Old Furnace State Park, Hopeville Pond State Park, Pachaug State Forest, Mashantucket Pequot Museum, Mansfield Hollow State Park, Bigelow Hollow State Park, Quaddick State Forest and State Park, Jerimoth Hill (RI).

The Route

Connecticut Route 169 (CT 169), designated a National Scenic Byway in 1996, runs across northeastern Connecticut, one of the least populated and rural areas in this largely urban state. The 32-mile drive explores colonial towns, passes whitewashed churches and old graveyards surrounded by stone walls, and takes in pastoral vistas with cornfields, horse pastures, and apple orchards. The drive runs south from the Massachusetts border and the town of North Woodstock to Lisbon and the road's junction with I-395 north of Norwich.

Connecticut Route 169

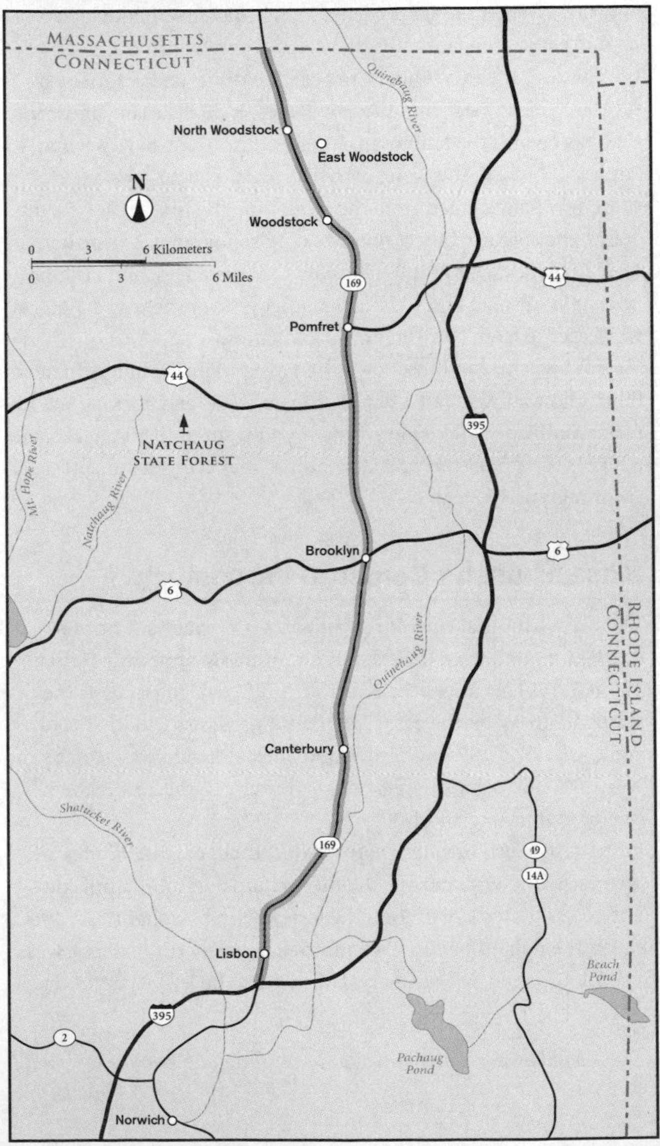

While Connecticut is one of the most citified states, with almost 90 percent of its residents living in metropolitan areas, its northeastern quadrant is the state's empty quarter, a region sometimes called the Quiet Corner. The drive, characterized by pleasant countryside and placid villages, is an intimate landscape that has been settled almost as long as Europeans have been in America. Indeed, this quiet corner of Connecticut appears in some ways unchanged from the landscape the first settlers in the 1680s knew. Don't expect the sweeping topography of northern New England, with grand mountains, deep valleys, and rushing rivers, but instead enjoy the area's quiet graciousness and a quick rural escape from busy highways and cities.

While you can buzz down the paved highway in less than an hour, allow at least half a day to stop, explore, and soak up the historical ambience. Take more time if you want to fish in a lake, take a hike, visit a museum or stately house, pick a bushel of apples, or enjoy a leisurely meal.

Massachusetts Border to Woodstock

The drive begins at the **Massachusetts–Connecticut border** south of Southbridge, Massachusetts. Here **Connecticut Route 169 (CT 169)** is easily accessed from I-395, which parallels the scenic highway on the east. The road dips across Muddy Brook and crosses CT 197 in North Woodstock at a four-way stop by the North Woodstock Congregational Church. To the east is the village of Quinebaug and I-395.

Drive south through North Woodstock, passing homes and then farmland with cornfields and orchards on rolling hillsides. The road crosses North Running Creek, climbs a slight hill, and enters the upland town of **Woodstock,** a lovely rural village with a spacious green. Woodstock Academy on the left occupies the

A shallow pond alongside Route 169 provides habitat for wildlife and waterfowl.

town's highest ground. The academy, founded in 1801, is the area's public high school, with over 1,100 students from Woodstock and surrounding communities including Eastford, Pomfret, Canterbury, and Brooklyn.

Woodstock

Woodstock, like other towns along the drive and elsewhere in upstate Connecticut, was founded by early Puritans from Massachusetts. The villages, like Woodstock, were built on hilltops with houses clustered around the town green or common and the meetinghouse, which is what Puritans called their churches. The hillsides below the towns were cleared for fields, with stones heaped in long fencerows along the edges.

Woodstock began as a humble "praying town," a village called Wabaquasset created in the mid-1600s by Puritan missionary John Eliot as a religious alternative for Native Americans. Here they could give up their religion, convert to Christianity, and pray for forgiveness and salvation. In 1675, however, King Philip's War started. Some natives sided with the English settlers and others with another Indian group led by a chief, King Philip. The town was deserted until 1682 when Massachusetts bought the region from the Mohegans. In 1686, 13 men moved down and refounded the town, naming it New Roxbury. The town's name was changed to Woodstock in 1690.

The **First Congregational Church,** established in 1690, rises on the southeast side of the green. Take a stop here to see the classic church and meander through its old burying ground, which boasts lots of historic graves including those of some Revolutionary War soldiers. If you're here for the night, stay at the **Inn at Woodstock Hill** in South Woodstock. The well-known inn overlooks the wooded hills west of the drive.

The First Congregational Church, established in 1690, rises along the green in Woodstock.

The **Woodstock Fair,** run by the Woodstock Agricultural Society in South Woodstock, is a popular town event that has been held every year on Labor Day weekend since 1860—one of the oldest continuous agricultural festivals in the United States. Folks from all over New England come for the festivities.

Roseland Cottage

Roseland Cottage, one of New England's unique houses, sits opposite the town green and the Congregational Church on the west side of CT 169. The cottage, if you could call a house that big a cottage, was built in a garish Gothic Revival style, with angular gables, trellis works, ornamental chimney pots, gingerbread trim, and fleur-de-lis crests, and then colored a joyful pink, in contrast with the more staid Yankee houses surrounding it. The house has been painted 13 colors, all shades of pink, and currently sports a salmon color.

Roseland Cottage, a National Historic Landmark, was designed by architect Joseph C. Wells for publisher Henry C. Bowen of Brooklyn, New York, as a summer home, a quiet refuge from the sweltering city. The house still houses Mr. Bowen's original furniture, wall hangings, and belongings. Bowen, a Woodstock native, entertained friends and business colleagues at the house, as well as four presidents—Ulysses S. Grant, Benjamin Hayes, Rutherford B. Hayes, and William McKinley—who spoke at his lavish Fourth of July celebrations. Grant, however, was the only sitting president to visit, and had to smoke his famous cigars out on the porch since Bowen didn't allow smoking and drinking in his home. Bowen, locally called "Mr. Fourth of July," organized huge celebrations that attracted over 10,000 people.

The cottage, open for tours, was named one of the 25 most important houses in America by *Fine Homebuilding* magazine in 2006. It's open from June 1 to October 15 from Wednesday to Sunday each week.

The historic Roseland Cottage in Woodstock is a wooden masterpiece painted joyful pink.

Pomfret

The drive rolls south from South Woodstock for 7 miles, passing farms, fields, and forest, and enters the town of **Pomfret,** which covers 40 square miles and includes several villages including Abington, Elliotts, Pomfret, Pomfret Center, and Pomfret Landing. Pomfret, named by English settlers after the town of Pontefract in Yorkshire, England, was incorporated in 1713. The original Indian name for the site was Mashamoquet, roughly translated as "great fishing place."

Pomfret sits astride the junction of CT 169 and US 44. On the northwest corner of the junction is the Vanilla Bean Restaurant, a popular sandwich shop famous not only for its terrific food but also as the place where Coca-Cola launched its short-lived beverage Vanilla Coke in 2002. If you keep alert, you might spot local celebrity, actress Renee Zellweger.

The 380-acre **Pomfret Street Historic District** on CT 169 or Pomfret Street is listed on the National Register of Historic Places.

Pomfret, settled in 1713, was named by colonists for their hometown Pontefract in England.

The listing notes, "No other town in the state contains such an impressive and cohesive ensemble of stylish summer houses . . ." The town also has other National Register properties including the Brayton Grist Mill; Israel Putnam Wolf Den; and the Pomfret Town House built in 1841.

Continue south for a mile from the CT 169–US 44 junction to the old town green spread across the flat rise of Pomfret Hill. The white First Congregational Church of Pomfret, organized in 1715, lifts its steeple over the eastern edge of the green on the east side of the highway.

Also on the east is the famous **Pomfret School.** A low stone wall separates the school's neat and orderly 500-acre campus from the asphalt. A long row of ivy-covered brick buildings house over 350 students, grades 9 through 12, at this college prep school founded in 1894 by William E. Peck. The campus was designed

Pomfret School boasts a neat 500-acre campus with ivy-covered brick buildings.

that year by famed landscape architect Frederick Law Olmsted, who also designed Central Park in New York City and Denver's City Park.

South of Pomfret Center and the school, US 44 takes leave and heads west. Turn here and drive a mile west on US 44 to 900-acre **Mashamoquet Brook State Park** for an historic stop or a longer leisurely visit. This parkland actually combines three state parks—the original Mashamoquet Brook, Wolf Den, and Saptree Run. The historic **Wolf Den** area was originally purchased in 1899 by the Daughters of the American Revolution. In 1742, Israel Putnam, later a Revolutionary War general, stole into the wolf den and killed what was supposedly Connecticut's last wolf, causing local sheep farmers to rejoice. The den is listed on the National Register of Historic Places. Nearby are a couple rock formations—Table Rock and Indian Chair. The park also has hiking trails, great fishing, swimming, and 2 campgrounds with 55 sites.

Brooklyn

Continue south from Pomfret on CT 169, passing a country junction with CT 101 at a traffic light, for 8 miles to **Brooklyn, Connecticut.** The drive passes spacious farms and orchards, where you can stop in the autumn and buy a bagful of crisp apples, before crossing the Brooklyn town line. Three miles later CT 169 intersects US 6 in the center of Brooklyn. This busy junction is the heart of historic Brooklyn.

The area around town was settled in the early 18th century after it was sold by the Mohegan Indians in 1703. It was part of Pomfret and Canterbury, while its northeastern sector was Mortlake, an estate established as a refuge for Puritans forced out of England by the restoration of the Stuarts. The town of Brooklyn was incorporated in 1786 and prospered as the seat of Windham County.

The **Brooklyn Green Historic District,** listed in 1982 on the National Register of Historic Places, is renowned for its diverse architecture, including houses and buildings built in Greek Revival, Colonial, and Federal styles. The center of the district and town is Brooklyn Green, an almost 2-acre square village green in the town center. Two streets, including CT 169, now slice across the green, dividing it into four triangles. Historic buildings on or around the green include the town hall, library, Trinity Episcopal Church, Federated Church of Christ, the Brooklyn Meeting House, and houses from the mid-1700s.

The **Brooklyn Fair,** established in 1809, is the oldest continuously operating agricultural fair in the United States. The fairgrounds are a half-mile south of the highway junction on CT 169.

Just past the drive's junction with US 6, next to the post office, is a statue of a gallant man on horseback. The statue honors one of the area's most famous heroes, Revolutionary War general Israel Putnam—remember him? The one who killed Connecticut's last lone wolf? After slaying the wolf, Putnam lived peacefully with his family on a farm in Woodstock, except for a stint as Major in the French and Indian War that lasted until April 20, 1775, when colonial patriots clashed with British redcoats at Lexington and Concord outside Boston. Putnam, already in the Brooklyn militia, heard the news and immediately rode 100 miles north in 8 hours to join the rebellion. Given the title General-in-Chief, Israel Putnam led his Connecticut troops at the Battle of Bunker Hill on June 17, when the rag-tag colonial army decisively defeated the British. Supposedly it was Putnam who said the infamous words, "Don't fire 'til you see the whites of their eyes." Two days later, General George Washington commissioned him as a Major General.

After Putnam's 1790 death, he was buried in Brooklyn's South Cemetery. His gravestone, however, was damaged by heavy visitation, so in 1888 his remains were removed and placed in a sarcophagus inside the monument below his statue. His epitaph on the monument reads: "Passenger, if thou art a soldier, drop a tear over the dust of a Hero, who ever attentive to the lives and happiness of his men dared to lead where any dared to follow."

Quinebaug River Valley

The drive runs south for 7 miles from Brooklyn to Canterbury, passing through deep woods and open fields with views across the gentle **Quinebaug River Valley** to the east. The byway traverses the rolling upland west of the river, a region designated the Quinebaug and Shetucket Rivers Valley National Heritage Corridor in 1995 and managed by the National Park Service in conjunction with local and state governments. The valley has been dubbed "the last green valley" in the Boston–Washington, DC, corridor. The mostly rural area includes 25 towns and a population of 300,000. It's called "a microcosm of the history of the nation, from the Native Americans and European settlement through its frontier days, the Industrial Revolution, and the many changes the 20th century has brought."

Halfway between the towns, the drive passes the **Moses Cleaveland Birthplace.** Cleaveland, a Revolutionary War soldier born here in 1754, was a shareholder in the Connecticut Land Company, which owned a large section of land in northwestern Ohio that was reserved for Connecticut. Cleaveland surveyed the area, called the Western Reserve, in 1796 and founded the city of Cleveland, Ohio, at a propitious location on the Cuyahoga River. The "a" was dropped from the city's name by a newspaper in the 1830s.

Canterbury & Prudence Crandall

The drive enters **Canterbury** and reaches a crossroads with CT 14 at town center. The town originally began as part of Plainfield, but separated and became Canterbury in 1703. It was named, of course, for the famous cathedral town of Kent in southeast England.

The Prudence Crandall Museum commemorates Connecticut's State Heroine and her fight against racism.

The **Prudence Crandall Museum,** on the southwest corner of the intersection of CT 149 and 14, is one of the most interesting historic sites along the scenic byway. Crandall was a Quaker schoolteacher and abolitionist who ran a boarding school in Canterbury between 1832 and 1834. In 1833, however, she admitted Sarah Harris, a local black girl, to the all-white school, becoming the first integrated school in the United States. In the resulting uproar, all the white girls left the school, so Prudence opened it as a school for African-American girls with 24 students, most being out-of-state boarding students from Boston and Philadelphia.

The school was vandalized by locals, who also refused to sell her supplies and groceries. By mid-1833, the bullying had not worked, so Canterbury residents took their case to the Connecticut government and the General Assembly passed the "Black Law," which forbade schools from accepting African-American students unless a town specifically allowed it. In 1834, the law was overturned by the courts, which led to a mob breaking all the school windows. The next day, Prudence, afraid for the lives of her students, closed down the school and left for Illinois and then Kansas. By 1886, Canterbury residents, so embarrassed by the bigotry, successfully petitioned the state to compensate Crandall for lost wages by awarding her a pension.

Outside the schoolhouse museum, a National Historic Landmark open for visitation, is a kiosk explaining the importance of Prudence Crandall and her stand for racial equality. The sign says: "The courageous stand against racism, sexism, and injustice taken by Prudence Crandall and her students helped make the rights and liberties expressed in the Declaration of Independence a reality for African Americans and for women of all races. That is their enduring legacy." For her efforts and inspiration, Prudence Crandall (1803–1890) is Connecticut's official state heroine.

The steepled First Congregational Church has been part of Canterbury life since its 1703 birth.

Lisbon & Newent

The last drive segment runs south for 8 miles from Canterbury to **Newent,** the town center of Lisbon, before reaching the drive's terminus at I-395. The highway continues south along the western edge of the broad Quinebaug River Valley. The river, with tree-lined banks, meanders in big wheeling turns beside broad cornfields.

Lisbon, a loose residential community, straddles CT 169. The center of town is the village of Newent, an old colonial hill town with a green lined with spreading maple trees. Newent's lovely steepled Congregational Church, organized in 1723 and built in 1858, lies on the east side of the highway. The **Newent Woods Heritage Trail** begins at the south side of the church and explores the surrounding woods, including an abandoned 19th-century farm with rusting implements and foundations made from heaped field stones, all overgrown with trees.

Across the street from the church is the old **John Bishop House,** built in 1810. The house, a local museum run by the Lisbon Historical Society, has 7 fireplaces and lots of unique features including a shaft that goes down to a well beneath the house. Another interesting historic site here is the **Anshei Israel Synagogue,** listed on the National Register of Historic Places, a mile east of the drive on CT 138. It's a small white country synagogue built in the Colonial Revival tradition. Also nearby is the historic **Norwich–Worcester Railroad Tunnel,** the first railroad tunnel in Connecticut and the oldest railroad tunnel still in use in the United States. The 300-foot-long tunnel was hewn through a rock bank along the Quinebaug River in 1837.

The drive ends a half-mile south of the church and John Bishop House at the junction of CT 169 and I-395. Hop on the southbound interstate lanes for a quick trip south to Norwich and the Connecticut coast drive. There is no entry ramp here for the northbound interstate lanes, so if you're heading north you'll have to reverse the drive to Canterbury and jog east 4 miles on CT 14 to I-395.

3 South Litchfield Hills

General description: This 58-mile loop drive travels across scenic hills and valleys, through historic Litchfield, then follows the Housatonic River Valley south to New Milford.

Special attractions: Litchfield, Kent, Kent Falls State Park, Housatonic River, Bull's Bridge, Macedonia Brook State Park, Sloane-Stanley Museum, Mount Tom State Park, White Memorial Conservation Center, Lake Waramaug State Park, Mohawk State Forest, Appalachian Trail, hiking, camping, fishing, canoeing, rock climbing, nature study.

Location: Northwestern Connecticut. The drive lies north of Danbury and New Milford in the southern Litchfield Hills.

Drive route numbers: US 202 and 7, CT 45, 63, and 4.

Travel season: Year-round. A good time to visit is May through October. Spring brings new greenery to the hillsides and flowering shrubs. Summers are pleasant, but can be hot. Autumn brings warm weather and spectacular fall colors. Winters are cold and snowy.

Camping: Macedonia Brook State Park, west of Kent, offers 51 sites April through September. Lake Waramaug State Park near New Preston has 77 sites April through September. Housatonic Meadows State Park at Cornwall Bridge (just north of the drive on US 7) offers 95 sites April through September. All are fee areas. Call the state parks at (860) 424-3200 or (866) 287-2757 for information and reservations. White Memorial Foundation has 2 campgrounds: Point Folly with 47 sites and Windmill Hill with 18 sites. Several private campgrounds are found along or just off the drive.

Services: All services in New Milford, Litchfield, and Goshen; limited services in other towns along the route.

Nearby attractions: Housatonic Meadows State Park, West Cornwall covered bridge, Housatonic State Forest, Bartholomew's Cobble, Berkshire Hills, Hartford attractions.

South Litchfield Hills

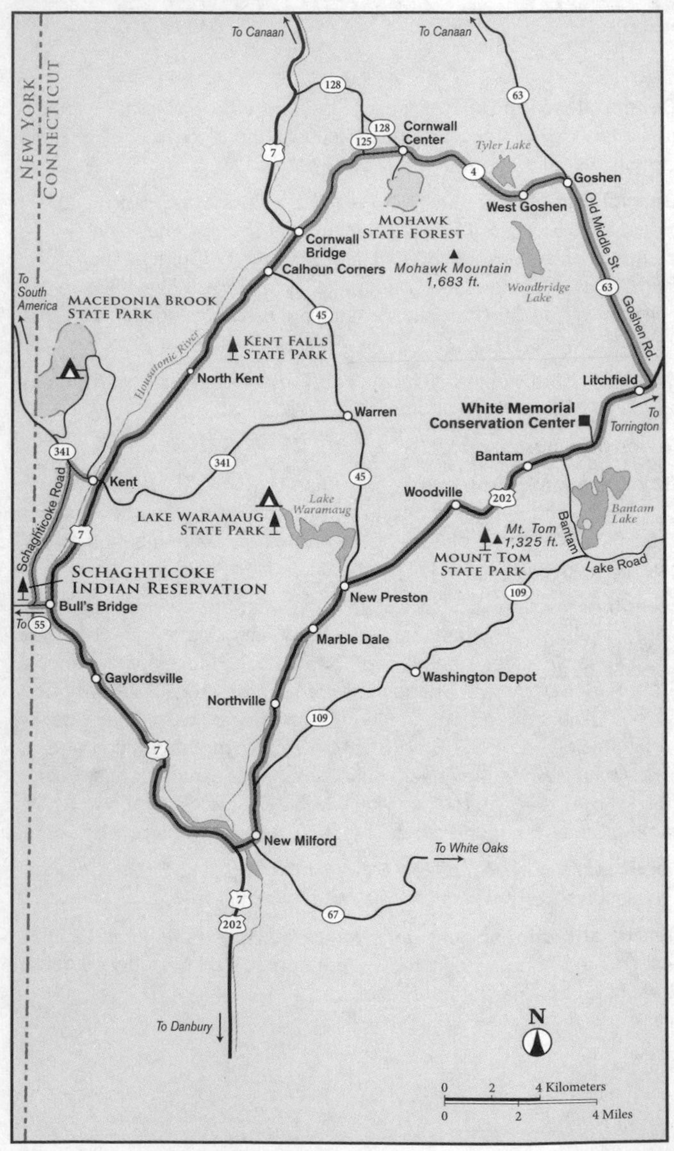

The Route

Sprawling across northwestern Connecticut, the Litchfield Hills form one of New England's most charming landscapes. The rounded, rolling hills are broken by sharp wooded vales and transected through their pastoral heart by the broad Housatonic River. Classic white saltbox farmhouses and red barns scatter across the hills, and old stone fences, some wildly overgrown, mark the edges of cleared fields and pastures. Amid this bucolic splendor sit some of New England's best-preserved 18th- and 19th-century villages, including Litchfield, Cornwall, and Kent. The South Litchfield Hills drive, a 58-mile loop that begins in New Milford, travels across rustic countryside, explores picturesque towns, and discovers natural wonders and hidden beauty.

New Milford to Lake Waramaug

The drive, readily accessible from the New York metro area and the Connecticut River Valley, begins at the junction of US 7 and 202 on the west bank of the Housatonic River in **New Milford,** 12 miles north of I-84's exit 7 in Danbury. Turn east (right), cross the river on a steel girder bridge, and enter the historic center of New Milford. This town, anchoring the southern end of the Litchfield Hills, is the largest in Litchfield County with a population of 28,000. Established in 1707, it was built on land bought from local Indians in 1703 by enterprising settlers from Milford on the south coast.

The expansive **New Milford Green,** one of the area's largest, runs a quarter-mile along Main Street. Grand Colonial homes bound the green, including the Town Hall. Its site marks the home of Roger Sherman, the only man to sign all four of America's founding documents—the 1774 Articles of Association, the 1776 Declaration of Independence, the 1777 Articles of Confederation, and the 1787 US Constitution. The **New Milford Historical Society Museum,** in an 18th-century house at the end of the green,

offers a glimpse into the town's past with historic portraits, furniture, china, and silver. Nearby are antiques shops, art galleries, and restaurants.

Follow US 202 through New Milford past old homes and bear northeast. The highway passes an old cemetery and quickly takes leave of the town as it passes into a broad, rolling valley studded with farms and houses. After a few miles the nonprofit **Henderson Cultural Center at Hunt Hill Farm,** an easily recognizable farm with a couple of houses, large barns, and twin silos, lies on the east side of the highway. The center offers changing monthly art exhibits and an acclaimed and popular cooking school. The highway next passes through Northville and Marble Dale, a pair of small towns with a handful of homes, and reaches the edge of New Preston after 8 miles.

Turn left on CT 45 here and drive up a hill into **New Preston** itself. This village offers numerous, excellent antiques shops along with an array of cafes, boutiques, galleries, and the New Preston Congregational Church. This delightful church with a tall, thin steeple was erected in 1853, although its congregation was established nearly a century earlier, in 1757. A red abandoned mill lies on the far side of town.

Continue up CT 45 to Lake Waramaug, the second largest natural lake in Connecticut; its Indian name supposedly means "place of good fishing." The lake is surrounded by several inns and 95-acre **Waramaug State Park.** On the lake's northwest corner, the park offers 77 campsites, boating, fishing, and a swimming beach. Nearby **Hopkins Vineyard** offers wine samples in an old barn and vineyard tours. The Litchfield Hills are an important Connecticut viticultural area. A few miles farther north on CT 45 lies the picturesque hamlet of Warren. This small town, named for General Joseph Warren, who was killed at the Battle of Bunker Hill, is dominated by its Congregational Church.

An overgrown stone wall lines a back road on the Schaghticoke Indian Reservation.

Mount Tom to Litchfield

Return south on CT 45 to US 202 to continue the drive. The highway runs northeast through Woodville and dips sharply into a wooded valley. It climbs steeply, swings across a hill, and after 14 miles reaches the entrance to **Mount Tom State Park.** The 232-acre state park, sitting on the east side of spring-fed Mount Tom Pond, includes a beach for swimming, water for nonmotorized boating, picnic tables alongside the lake and in the woods above, and an excellent mile-long trail that winds through dense woods to a 3-story lookout tower atop 1,325-foot-high Mount Tom. The lookout, perched above the tall trees below, yields a superlative view of the Litchfield Hills. The forested hills roll away below the peak, their green flanks broken by grass and the glint of occasional lakes. Mount Frissel, with Connecticut's highest point on its ridge at 2,379 feet, rises to the northwest in Massachusetts, while the white church spires of Litchfield lift above the green hills on the northeast horizon.

US 202 skirts the west edge of Mount Tom Pond and shortly afterward reaches Bantam. This small town is bordered by the **White Memorial Foundation,** a unique woodland reserve that protects more than 4,000 acres. Turn off US 202 at the woodland reserve's sign and follow a gravel road 0.5 mile to a parking area at the museum and gift shop. This privately administered preserve, the state's largest wildlife refuge, was established by a brother and sister of a wealthy family that summered in nearby Litchfield in the late 19th century. Remembering their formative summers here, Alain and May White acquired the land between 1908 and 1912 and set up a land trust in memory of their parents. The foundation preserves the area's unique woodlands and wetlands, providing environmental education, recreational opportunities, and research facilities.

More than 35 miles of nature trails lace the parkland, winding through woodlands of black cherry, Canadian hemlock, white oak, and large white pines. Other trails explore swamps and peatlands on boardwalks, passing characteristic marsh plants that include skunk cabbage, bulrush, turtlehead, and ferns, and provide great birding opportunities. The trails are also used for horseback riding and, in winter, cross-country skiing. Boating and fishing are popular pastimes along the Bantam River and Bantam Lake here, where a campground is available for overnight visitors. Inquire at the visitor center for the current schedule of nature programs, field trips, and lectures.

Litchfield

The highway rolls through Bantam past antiques shops and reaches Litchfield in a scant 3 miles. **Litchfield,** considered one of the best-preserved late-18th-century New England villages, makes a cluster around the expansive town green. The town is split by two principal residential thoroughfares, named, with Yankee brevity, North and South Streets. William Adam described the streets in 1897: North Street "is a magnificent way, broad and straight, with ample plats of grass, bordered by fine old houses with spacious yards, the ideal of a New England street, while its southern continuation curves gently past houses of much the same sort and once the homes of distinguished men and women." The streets haven't changed much in the last 100 years.

Sprawling across a rounded hilltop some 1,200 feet above sea level, Litchfield is part of old New England and abounds in historical nuances. The town, named by its first English settlers for the old country's Lichfield, was settled in 1720. As surrounding hills were cleared and farmed, Litchfield grew into the county seat by 1751. The village thrived in the late 18th century as a freighting and passenger crossroads between the upper New England cities and New York and the Hudson River settlements.

The American Revolution thrust Litchfield into an infant nation's arms. Its relative isolation made Litchfield an ideal place to confine royalist prisoners, including the royalist mayor of New York, David Matthews, and William Franklin, the royal governor of New Jersey. The town's advantageous location on the main highway between Hartford and New York made it an ideal supply depot and storehouse for George Washington's Continental Army. A statue of despised British King George III was shipped here from New York in 1776 and melted into 42,088 bullets to help expel the British soldiers from the new nation. The war gave Litchfield a decidedly military atmosphere with numerous generals, including Washington and Lafayette, visiting the town.

Litchfield continued to prosper after the revolution and by 1810 was Connecticut's fourth-largest municipality. Industry and manufacturing fueled this now-quiet town, with 18 sawmills, forges, nail and comb factories, a paper mill, and 5 tanneries. The town's boots and shoes were renowned for their quality and superb workmanship. By the 1830s, however, Litchfield's fortunes began to decline as the Industrial Revolution's burgeoning web of railroads bypassed the hill-bound town and spread its industry into the Connecticut Valley and along rivers that provided hydropower. This decline was a blessing in disguise, preserving Litchfield's historical treasures and ambience for late-20th-century historians and travelers.

The three-centuries-old town green is the first of Litchfield's historic sites encountered along US 202. The long, undulating green, studded with benches and sidewalks and shaded by towering trees, was created in 1723 at the town center. The rest of Litchfield, a historic district composed of almost 500 buildings and listed on the National Register of Historic Places, surrounds the green and the residential side streets. Begin a good walking tour of old Litchfield by parking near South Street and the green.

The 1721 First Congregational Church, the town's most noted landmark, soars to the north above the green and its trees with a white facade, bell tower, and conical steeple. Rumor says this is the most photographed church in New England. Nearby is the **Litchfield History Museum,** housed in a brick library that dates from 1900. The museum's galleries offer portraits of early

Litchfield luminaries by 18th-century painter Ralph Earl, early American toys, furniture displays, and exhibits detailing the area's history. Across the street from the museum is a narrow cobblestone alley that leads to Cobble Courtyard, a brick 19th-century livery stable that now houses shops.

A block down South Street sits the **Tapping Reeve House and Law School.** The nation's first law school, it was established by Tapping Reeve in 1775. After several years of teaching in his house, Reeve moved the school in 1784 into a one-room building next door. The small, unheated school, so cold in winter that ink froze in the wells, taught law to some of early America's finest legal and political minds. Graduates included Reeve's brother-in-law and vice president Aaron Burr, educator Horace Mann, western painter George Catlin, inventor Sydney Morse, and statesman John C. Calhoun, as well as 3 Supreme Court justices, 2 vice presidents, 26 senators, 6 cabinet members, more than a hundred members of Congress, and 16 governors, including 6 of Connecticut's. The school closed in 1883.

After a stroll around the green and the surrounding spacious streets, turn north onto CT 63 on the north side of the green. The highway, following North Street, passes numerous historic homes, notably the **Pierce Academy** and the **Beecher House.** The Sarah Pierce Litchfield Female Academy was established in 1792 to cultivate the intellectual and cultural potential of young American women. More than 3,000 students enrolled in the school during its 41-year history. The **Beecher House** was home to Litchfield's most prominent family. Pastor Lyman Beecher landed here as a country preacher in 1810. His offspring became two of the 19th century's most noted and influential Americans—Reverend Henry Ward Beecher and Harriet Beecher Stowe, author of the anti-slavery novel *Uncle Tom's Cabin.* Other famed Litchfield residents include Revolutionary War officer Benjamin Tallmadge, patriot Nathan Hale, the Wolcott family, and a prolific Mary Buel, "wife of Dea. John Buel, Esq." Her 1768 gravestone notes her death at age 90 after "having 13 children, 101 grandchildren, 247 great-grandchildren, and 49 great-great-grandchildren; total 410. Three hundred and thirty-six survived her."

The drive's next leg runs northwest to Goshen along CT 63. The highway quickly leaves Litchfield and passes forests, pastures, and occasional houses tucked among trees. In Goshen the highway intersects CT 4 at a small rotary. Turn west (left) on CT 4 at the rotary.

Goshen owns a large fairground, which hosts the annual Connecticut Agricultural Fair in late July and a Scottish festival in October. Another well-tended colonial village, Goshen preserves its history at the **Goshen Historical Society Museum,** a white colonial building on CT 4.

Outside Goshen the road passes a cemetery and reaches West Goshen after 2 miles. The highway then rolls west past Tyler Lake, through deep woods, and after a few miles reaches the turnoff to Mohawk State Forest. Turn south onto Allyn Road and go 1.6 miles to the forest parking area and headquarters.

Mohawk State Forest

Mohawk State Forest, one of Connecticut's best woodland preserves, spreads across the slopes of 1,683-foot Mohawk Mountain. The 3,351-acre parkland protects some unique ecosystems, including a 2-acre black spruce bog. Although similar bogs are common in northern New England, they are rarities this far south. A short boardwalk trail explores the bog, initially passing through a forest of tall red pines before reaching the bog's edge. The dominant tree in the bog is black spruce, a boreal tree that reaches heights of 30 feet. These stunted, spindly trees are well adapted to the bog's harsh and acidic peat, a mass of undecayed organic material deposited in the bog. Sphagnum moss is another common plant found here along with sheep laurel, dwarf huckleberry, and sundew, a carnivorous plant that digests insects trapped in sticky hairs on its round leaves. Other trails venture through the forest, climbing to lookouts and winding through dense woods.

Mohawk Ski Area, also in the state forest, is Connecticut's largest ski and snowboard area. It offers 25 groomed trails and slopes, 7 ski lifts, snowmaking equipment, night skiing, and a lodge.

Past the state forest turnoff, the highway drops steeply downhill and reaches the junction of CT 4, 128, and 43. Keep left on CT 4. The road passes the turnoff to Mohawk Ski Area and drops down another hill to **Cornwall Center,** a township established in 1740. All the Cornwalls in the vicinity are somewhat confusing—Cornwall, Cornwall Bridge, Cornwall Center, Cornwall Hollow, West Cornwall, and East Cornwall. The area's famed Cornwall covered bridge lies to the north in West Cornwall, not in Cornwall Bridge. (See the North Litchfield Hills Scenic Route.)

A good Cornwall side trip ventures to the famous **Cathedral Pines Preserve.** Turn left on Pine Street at the junction of CT 4 and CT 125 in Cornwall. The road leads to the Cathedral Pines, a once-immense old-growth white pine forest that was decimated by a rare tornado on July 10, 1989. The stand of 200-year-old pines, one of the few in all New England untouched by 19th-century land clearing and charcoal cutting, was a stunning sight before its destruction. The Nature Conservancy, which manages the 42-acre property, has allowed nature to take its course and has left the mangled trees as testimony to nature's wild caprices.

Along the Housatonic River

The drive continues southwest from Cornwall on CT 4 for 4 miles to **Cornwall Bridge,** the Housatonic River Valley, and a junction with US 7. Turn south (left) on US 7 toward New Milford for the last 23-mile segment of the drive. At Swift Bridge, 1 mile south, US 7 intersects CT 45. This 11-mile road, running southeast to New Preston, makes a good side excursion. It climbs over hills and dips through broad valleys broken by woods, farms, and stone fences. The drive, however, runs southwestward along the eastern edge of the broad **Housatonic River Valley** on a designated Connecticut Scenic Road.

The **Housatonic River,** paralleling the drive from here to New Milford, is a part-wild/part-workhorse river that once powered industry along its banks. The river, with an Indian name meaning "place beyond the mountains," originates in the Berkshire Hills and wanders 132 miles southward across Connecticut before emptying into Long Island Sound.

Four miles from Cornwall Bridge, the drive reaches **Kent Falls State Park.** This idyllic spot, popular with picnickers and hikers, is a broad, grassy expanse framed by thick woods and the 250-foot-high cascade Kent Falls. Kent Falls Brook, fed by springs high in the Litchfield Hills above the valley, riffles and tumbles over a series of rocky steps. A 0.25-mile trail climbs 250 feet from the falls' base to a higher viewpoint. The area also offers picnic tables and restrooms.

The highway runs along the valley edge past houses, occasional farms, and an old cemetery. **St. John's Ledges,** accessed via the famed Appalachian Trail, are on a scruffy cliff poised 500 feet above the rural valley. The cliff, composed of ancient metamorphic gneiss deposited billions of years ago as volcanic debris, yields superb views of the valley from its summit as well as slab routes for rock climbers. The cliff and trail are accessed from Skiff Mountain Road on the west side of the river.

Flanders Cemetery, south of the falls, offers a clean vista of St. John's Ledges and the broad river valley. The cemetery is worth a look too. The grave of 39-year-old Captain Jirah Smith, who was killed in the Revolution, boasts the inscription: "I in the Prime of Life must quit the Stage, Nor see the End of all the Britains Rage."

South on US 7 is the **Flanders Historic District,** a cluster of 18th-century homes that are part of the original Kent. One home, now the Flander Arms bed-and-breakfast, dates from 1738.

Kent Falls cascades 250 feet down a series of leaps at popular Kent Falls State Park.

Across the highway is **Seven Hearths,** the former home and studio of portrait artist George Laurence Nelson. The 1754 house, operated by the Kent Historical Society, displays Nelson's prolific works and is open for tours on weekends in July and August.

Kent to New Milford

The **Sloane-Stanley Museum,** on the northern outskirts of **Kent,** lies west of the highway along the riverbank. This interesting museum houses a large collection of early American tools and implements collected by Eric Sloane, an artist and writer who resided in nearby Warren. The museum was established in 1969 after Sloane donated his collection to the building erected by Donald Davis, president of the tool-making Stanley Works on the site of the old Kent Iron Furnace. The collection includes an astonishing number of ingenious wooden and iron tools fashioned by early settlers to eke a life out of New England's wilds. Excellent interpretative displays illustrate the uses of these arcane, unrecognizable tools. Sloane's studio and workshop, recreated from his home, are also displayed along with a furnished replica of an early American house and an 1826 pig iron furnace. After his death in 1985, Sloane was buried under a spreading maple on the property.

Kent, founded in 1731 and once a bustling iron center, is a charming old town reputed for its wide selection of antiques shops, galleries, and craft stores. Its streets are lined with fine homes, shops, and a white, turreted First Congregational Church established in 1741 and built in 1849. Turn right in the town center on CT 341 to visit **Macedonia Brook State Park** and St. John's Ledges. The state park offers 51 campsites, fishing, and hiking trails.

The highway continues south from Kent across the green valley floor. Occasional views open onto the mirrorlike Housatonic

Flanders Cemetery, south of Kent Falls, has numerous gravestones that date back to the American Revolution in the 1770s.

River and green pastures and farms. **Bull's Bridge** is reached after 4 miles. Turn west to its covered bridge, dam, and an abrupt rocky gorge with waterfalls and frothy rapids. The bridge, one of two remaining covered bridges of the 18 that once spanned the Housatonic, was originally built in 1842 for $3,000. The bridge was used on a route that hauled iron products from Jacob Bull's furnace to New York markets. The bridge was later raised 20 feet after the Connecticut Power & Light Company built a power plant at the site. Just west of the bridge and river is a parking area for hikers, fishermen, and picnickers. The Appalachian Trail travels north from here along the river and over St. John's Ledges en route to its Maine terminus.

An excellent side-trip runs north from here on Schaghticoke Road on the west bank of the river. The road narrows to rough one-lane pavement and gravel as it winds through tall sycamores along the narrow west bank of the Housatonic River. The **Schaghticoke Indian Reservation,** the remnant of a once-great Indian nation, huddles along the road. The old burying ground sits alongside the road among shady trees. One stone is engraved "Eunice Mauwee, a Christian Indian, 1756–1860." Farther north the road passes Kent School, a private academy, before rejoining CT 341 just west of Kent.

The drive continues by heading back south on CT 341 to Bull's Bridge. From here the highway rolls south along the river's east bank for 7 miles to **Gaylordsville,** a pleasant riverside town founded in 1725. Nearby sits the last one-room schoolhouse in Connecticut. It opened in 1765 and held classes for two centuries until closing in 1967. The road swings over the river and meanders another 7 miles south through quiet, wooded countryside broken by glimpses of the river and secluded homes. Past a power plant, the drive ends at the girder bridge at the drive's starting point and the busy junction of US 7 and US 202 in New Milford.

The gleaming white Congregational Church in Kent was built in 1849.

4 North Litchfield Hills

General description: A 56-mile loop drive that wanders across wooded hills and up the Housatonic River in northwestern Connecticut's scenic Litchfield Hills.

Special attractions: Housatonic Meadows State Park, West Cornwall Covered Bridge, Housatonic State Forest, Appalachian Trail, Norfolk, Yale School of Music music festival, hiking, camping, fishing, canoeing, nature study.

Location: Northwestern Connecticut.

Drive route numbers: US 7 and 44, CT 41, 272, and 4.

Travel season: Year-round. May through October is a good time to visit. Summers are pleasant, but can be hot. Autumn brings warm weather and spectacular fall colors. Winters are cold and snowy.

Camping: Macedonia Brook State Park, west of Kent and just south of the drive, offers 51 sites April through September. Lake Waramaug State Park near New Preston south of the drive has 78 sites from April through September. Housatonic Meadows State Park at Cornwall Bridge on US 7 offers 95 sites April through September. Several private campgrounds are found along or just off the drive.

Services: All services in most towns along the drive, including Winsted, Canaan, Sharon, and Cornwall Bridge.

Nearby attractions: Sloane-Stanley Museum, Kent Falls State Park, Litchfield Historic District, Mount Tom State Park, Lake Waramaug State Park, Mohawk State Forest, Mohawk Ski Area, White Memorial Conservation Center, Bartholomew's Cobble, Berkshire Hills.

The Route

The North Litchfield Hills drive explores the rural northwestern corner of Connecticut, traversing west from Winsted across rolling hills studded with occasional towns before turning south along the New York border and driving up the narrow valley of the Housatonic River to end at Canaan just south of the Massachusetts border.

North Litchfield Hills

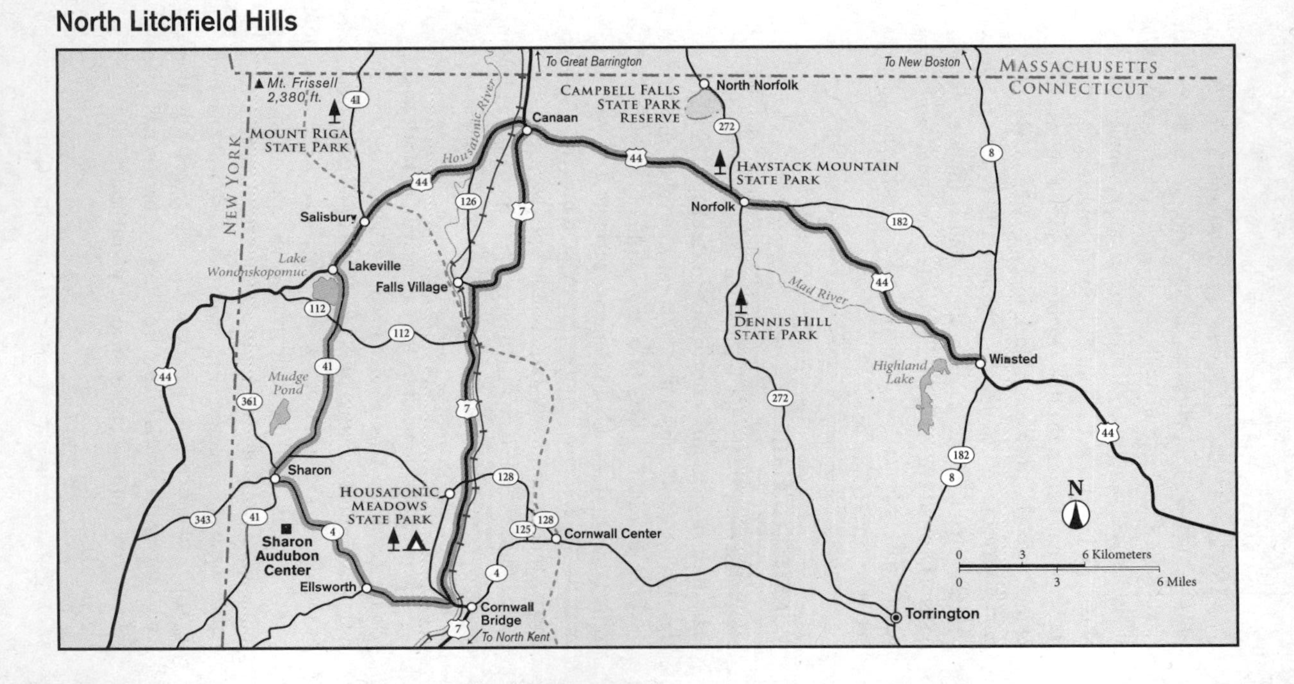

The drive begins on the west side of Winsted at the junction of US 44 and CT 183. **Winsted,** the second-largest town in Litchfield County, is a busy old mill town that reflects its working-class roots in stark contrast to surrounding rural communities. Lying almost midway between Winchester and Barkhamsted, Winsted draws its name from the first syllable of one and the last syllable of the other. The town was a clock-making center during the early 19th century. Later, during the Industrial Age, its factories manufactured wire, electrical products, and 12 million stickpins a day. The town offers a good selection of restaurants and shops, retaining its industrial heritage with old brick factories lining its congested streets.

Winsted to Canaan

Head west on US 44 from Winsted to take the drive's first leg, which runs 16 miles northwest to **Canaan.** The highway leaves Winsted, climbs out of its sprawling valley, and enters the Litchfield Hills. These hills, the southern end of the Green Mountains–Berkshire Hills range in Vermont and Massachusetts, are composed of a rolling upland broken by valleys. After 1 mile the highway passes Mad River Dam, and after 6 more miles enters Norfolk. This stretch of road offers pleasant, wooded country interrupted by small hill farms and occasional open pastures with good views to the south.

Compared to some of its Puritan neighbors, **Norfolk** is a relative latecomer. This isolated town, spread across stony, rolling hills, was originally settled in 1744 and incorporated 14 years later by the town's 44 voters. Norfolk sat along the old Albany–Hartford turnpike, a busy overland route that carried coastal goods inland to the upper Hudson River frontier. When the American Revolution began at Lexington in Massachusetts, 24 local men took up arms and marched to Boston to begin the fight for freedom. After the war the town's hilltop isolation and hard-scrabble terrain kept Norfolk from becoming an agricultural center. Instead the townspeople built an industrial economic base in the early 1800s, relying on nearby forests for lumber and the Blackberry River for hydroelectric power.

When its manufacturing era declined, Norfolk used its country charm to flourish as a summer resort for the well-heeled. The town's wealthy ambience stems from this period in the late 19th century. Immense homes, many of them now elegant bed-and-breakfasts, line Norfolk's streets. Interesting buildings include the handsome **Norfolk Library,** built in 1889 by Isabella Eldridge, and the **Norfolk Historical Society,** housed in the 1840 Norfolk Academy on Main Street, which has displays of local relics and exhibits about three prominent Norfolk families, the Battells, Eldridges, and Stoeckels. The Eldridges built the library and also commissioned Stanford White to design a three-tiered fountain with watering troughs for dogs, horses, and people near the village green. The Battell and Stoeckel families established a lasting legacy in Norfolk by leaving their rich estates to the Yale School of Music. The town hosts a renowned summer music festival in the school's music shed.

US 44 intersects with CT 272 in Norfolk's center. A short run south on CT 272 leads to **Dennis Hill State Park,** a small parkland that offers room for picnicking and walking. The area's centerpiece is a stone observation pavilion atop Dennis Hill, providing spacious views of the surrounding hills.

A mile north from the highway junction on CT 272 is **Haystack Mountain State Park.** This scenic area has an excellent hiking trail that winds through shady woods to the summit of 1,716-foot Haystack Mountain. A round stone tower with a spiral staircase offers spectacular vistas above the treetops of Connecticut's Litchfield Hills, the Berkshire Hills in Massachusetts to the north, and the Taconic Mountains in New York to the west. This long, thin range of mountains straddles New York's eastern border from here north to central Vermont.

The drive passes a third state park, **Campbell Falls State Park Reserve,** another 6 miles north of Norfolk off CT 272. Lying near the Massachusetts border, it is named for a series of cascades that tumble over rock ledges here. The park includes opportunities for hiking, picnicking, nature study, and photography, but has no developed facilities.

Crossing the divide that separates the Connecticut and Housatonic River drainages just west of Norfolk, the drive continues northwest along US 44 for 7 miles from Norfolk to Canaan. After crossing this nondescript divide, the highway drops into the broad valley of the westward-flowing **Blackberry River.** The river swings in wide meanders across the valley floor, its banks lined with black locust and cottonwood, a poplar more commonly found in the Midwest. High mountain ridges hem the valley to the south, forming a long, forested wall. Occasional farms and homes dot the pastoral scene.

Canaan

The blacktop runs through East Canaan and, 2 miles later, enters the village of **Canaan.** A crossroads town in the wide Housatonic River Valley at the junction of US 44 and 7, Canaan sits a scant 2 miles south of the Massachusetts state line. Like other northwestern Connecticut villages, it thrived with the area's early iron industry. The remains of abandoned iron forges are scattered throughout the Canaan area; most, however, are on private property. The **Beckley Iron Furnace,** listed on the National Register of Historic Places, is the best-preserved site to visit. Reach it by turning from US 44 onto Furnace Hill Road east of town. Head right at the stop sign and drop to the furnace ruins along the banks of the Blackberry River. East of the turn on US 44 is **Lone Oak Campsites,** one of the state's largest private campgrounds with more than 500 sites.

The **Canaan Union Station,** built in 1872 and listed on the National Register of Historic Places, sits in the center of town. The yellow station, one of the oldest in continuous use, is home to the Housatonic Railroad. The railroad runs excursion trains south to West Cornwall alongside the Housatonic River.

Salisbury, Lakeville, & Sharon

From here, the drive makes a 40-mile loop southwest of Canaan along US 44 and CT 41 and 4 to Cornwall Bridge, returning on US 7 along the Housatonic River. Head southwest out of Canaan on US 44. The highway leaves town, crosses the Blackberry River, and after a mile reaches the Housatonic. The highway is a designated Connecticut Scenic Highway for the next 8 miles, as the tree-lined river, descending gently southward, slowly unwinds across the broad valley. After another mile the road crosses the

The one-lane West Cornwall Covered Bridge, built in 1837, spans the Housatonic River.

river, works west over a ridge, passes private Salisbury School, and tilts down into a wide valley and Salisbury.

Salisbury, founded in 1741, and neighboring Lakeville, 2 miles south, were major iron ore centers in the 18th century. The almost pure iron smelted and forged in these communities proved crucial to the colonial effort in the American Revolution, supplying cannons, shot, and balls to George Washington's Continental Army. The iron industry here was so important to the Revolution that the ironworkers were supplied with meat and plentiful rum to sustain their work. The town's iron industry also produced war materiel during the War of 1812, and later provided iron used in the growing rail network that spread across the United States in the 19th century. Salisbury is now a quiet community dominated by a white Congregational Church built in 1800 and the stone Scoville Memorial Library. The Appalachian Trail, the long-distance footpath from Georgia to Maine, threads across the north side of town before heading into the Berkshire Hills to the north.

To take a good side trip, travel north on CT 41 for a few miles to the Massachusetts border. The road traverses a countryside of fields and occasional grassy pastures framed by dense woodlands. The **Under Mountain Hiking Trail** is reached 3 miles north of the junction of US 44 and CT 41. This trail meanders west to the top of 2,323-foot Bear Mountain, the highest mountain summit in Connecticut. The state's high point, however, is at 2,380 feet on the south side of Mount Frissell a mile to the west. Frissell's summit is in Massachusetts.

Back on the main scenic route, **Lakeville** sits 2 miles southwest of Salisbury. This once-flourishing iron town is now a peaceful rural community dotted with spacious colonial homes, including the **Holley-Williams House.** The house, built in 1768 and enlarged in 1808 by ironman John Milton Holley, was inhabited by five generations of the Holley family until 1971 when it

Wild lupines surround old wagon wheels in an abandoned field in the Litchfield Hills.

was deeded to the town. This unique house museum offers a glimpse into daily 19th-century life with period furniture and portraits, as well as an iron furnace model and tool displays in the adjoining Carriage House. One of Lakeville's most famous residents was patriot Ethan Allen, who lived here before his Green Mountain Boys captured Fort Ticonderoga from the British in May 1775. Allen was an owner of the Lakeville Forge between 1762 and 1765. The forge became a cutlery factory in the mid-1800s. Lakeville was also renowned for the excellent knives produced by the Holley Manufacturing Company in Pocketknife Square.

In Lakeville, turn south (left) onto CT 41, then drive 7 miles south to Sharon. The road runs past Lake Wononskopomuc and the Hotchkiss preparatory school grounds before entering a mixed woodland of pine and hardwoods interrupted by occasional marshy fens, dairy farms, and colonial houses. You'll get good views of the rolling hills from fields and pastures east of the highway. South of the highway's junction with CT 112, the blacktop undulates along a broad ridge crest. Mudge Pond, a long glimmering lake, nestles in a valley below.

The village of **Sharon,** like so many other Litchfield towns, is quintessential New England with a picturesque village green dating from 1739, a spired white Congregational Church, and broad streets lined with handsome homes. Town attractions include the **Gay-Hoyt House,** where the Sharon Historical Society displays local artifacts and paintings. Sharon's landmark is the 1884 cut-stone Hotchkiss Clock Tower on the south side of the green.

At the junction of CT 343, 41, and 4, continue southeast on CT 4 for 8 miles to Cornwall Bridge. The highway climbs away from the town into the hills to reach the **Sharon Audubon Center and the Emily Winthrop Miles Wildlife Sanctuary,** sprawling across the valley floor, in a couple of miles. This 1,147-acre nature

A marshy lake, providing ideal wildlife habitat, reflects the morning sky along the scenic route.

preserve offers almost 12 miles of hiking trails that explore surrounding forests and meadows, as well as a children's adventure center, a bookstore, and nature exhibits.

The road ascends the broad crest of the hills and rolls southeast past old farmhouses dating from the 1700s and overgrown stone fences dividing pastures. After a few miles it twists into a narrow canyon, passes a small roadside picnic area, and weaves steeply downhill through dense woods to the Housatonic River Valley and the junction with US 7 at Cornwall Bridge. A wealth of Cornwall-named locations are found in the immediate vicinity—Cornwall, West Cornwall, East Cornwall, and Cornwall Bridge. The area's covered bridge is not found in the town of Cornwall Bridge, but farther north on the drive in West Cornwall.

Up the Housatonic River

Turn north (left) on US 7 for the drive's last 17 miles back to Canaan. The highway runs north along the narrow west bank of the Housatonic River through a spectacular canyon lined with steep, wooded slopes broken by occasional rock steps. **Housatonic Meadows State Park,** a 451-acre area, straddles the river for the first few miles. It offers excellent fly fishing for trout, picnic tables for lunch, the 2.5-mile Pine Knob Loop Trail, which climbs to a 1,160-foot viewpoint, and a 95-site campground. Canoes and kayaks are available for rental from a roadside outfitter to anyone willing to brave the river riffles.

The canyon narrows past the campground and 0.5 mile later reaches **West Cornwall** and its long covered bridge. The village on the river's east bank is reached via the sturdy, one-lane, red covered bridge built in 1837. Clatter across the bridge and park in the narrow streets of West Cornwall. This picturesque village offers several interesting craft shops, including a pottery shop and furniture maker.

The drive continues north along the deep canyon floor, passing dense woods in 9,543-acre **Housatonic State Forest.** The forest, covering steep hillsides on both sides of the river, is broken into a dozen tracts of wooded land. The valley opens and widens as the road runs north. A few miles past the highway's intersection with CT 112 and the Appalachian Trail, the road crosses the river and enters **Falls Village.** Named for the nearby Great Falls on the Housatonic River, Falls Village was once part of Canaan. The town, like its neighbors, prospered with the iron industry, along with a busy lumber mill and hydroelectric dam. The frothy **Great Falls** are easily viewed from overlooks reached from the village center.

The drive continues north on US 7 from Falls Village, following a broadening river valley for 6 miles to the route's terminus at the junction of US 7 and 44 in Canaan. This segment traverses a lovely valley flanked by lofty ridges to the east and dense thickets of trees and shrubs that choke the placid river's banks. Near Canaan, the highway crosses the Blackberry River and enters a residential area before this drive ends at US 44.

5 Rhode Island Coast

General description: A 61-mile drive along Rhode Island's coast and Narragansett Bay between Westerly and Newport.

Special attractions: Watch Hill, Misquamicut State Beach, Burlingame State Park, Ninigret National Wildlife Refuge, Ninigret Conservation Area, Fort Ninigret, Charlestown, Trustom Pond National Wildlife Refuge, Point Judith, Silas Casey Farm, Gilbert Stuart Birthplace, Conanicut Island, Fort Wetherill State Park, Newport attractions, Cliff Walk, scenic views, hiking, birding, fishing, historic sites.

Location: Southern Rhode Island.

Drive route numbers: US 1 and 1A, RI 108 and 138.

Travel season: Year-round. The weather along the coast is generally very pleasant. Warm summer temperatures are moderated by sea breezes.

Camping: Burlingame State Park north of US 1 has 755 campsites with water, washrooms, showers, toilets, and picnic tables. Charlestown Breachway offers 75 sites for self-contained units. Fisherman's Memorial State Park at Galilee has 147 trailer/RV sites and 35 tent sites.

Services: All services in Westerly, Watch Hill, Charlestown, Wakefield, Narragansett Pier.

Nearby attractions: Providence attractions, Lincoln Woods State Park, Connecticut coastline, Mystic Seaport, Cape Cod National Seashore.

The Route

This 61-mile excursion begins in Westerly on the Connecticut border and roughly follows the Rhode Island shoreline east to Narragansett. Here it works north along Narragansett Bay, crosses the bay to Conanicut Island, and finishes in the famed city of Newport at the southern tip of slender Aquidneck Island. The drive offers pleasant scenery, historic sites and mansions, dense woodlands,

Rhode Island Coast

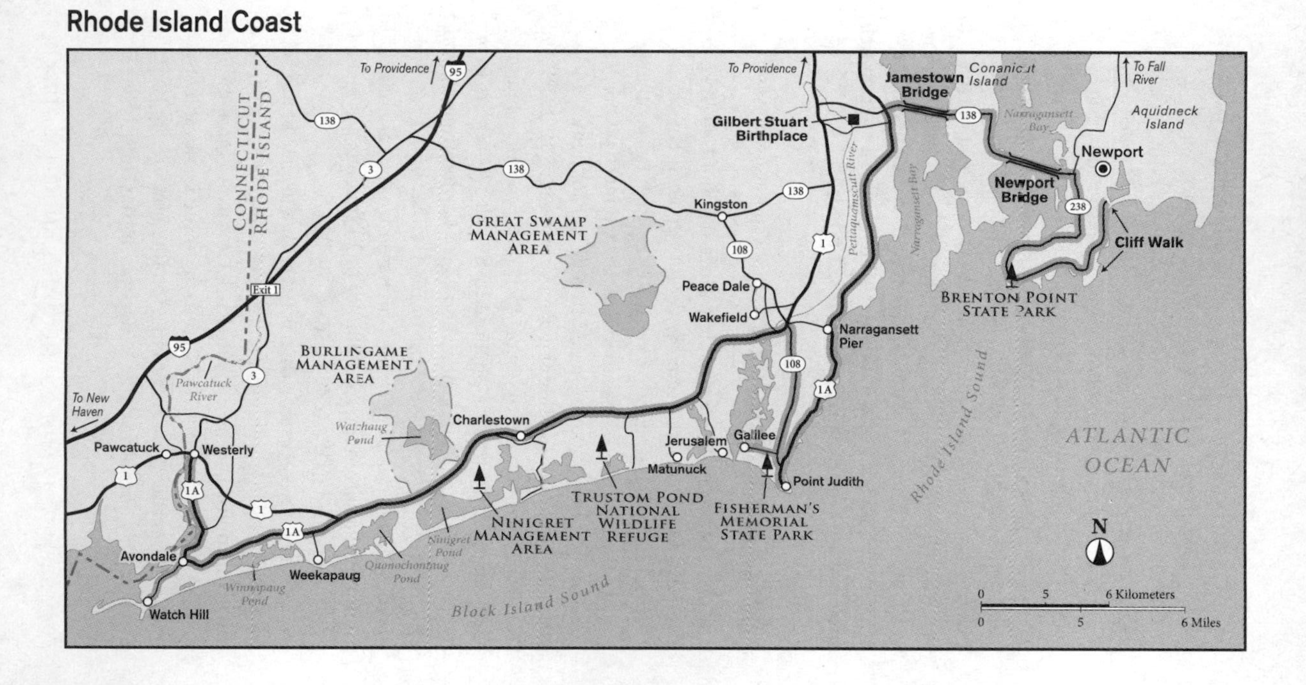

rock-rimmed peninsulas, long sand spits and beaches, salt marshes, and a maze of inlets, coves, and harbors.

Rhode Island, nicknamed the Ocean State, is the nation's smallest state, measuring only 48 miles long by 37 miles wide. More than 20 percent of Rhode Island is consumed by Narragansett Bay, a wide slice of water reaching across the state. The bay encompasses 12 miles of water on the south, narrows to 3 miles in the north, and extends 28 miles inland. Three major islands—Aquidneck, Conanicut, and Prudence—split the bay into two rough halves. The twisting shoreline of the bay, which is the state's premier natural feature, totals more than 300 miles. The oceanfront between Point Judith and Watch Hill is a long, straight stretch of sand beaches, lagoons, and salt ponds.

Westerly to Watch Hill

The drive begins in **Westerly,** a few miles south of I-95. Take exit 1 and drive south on RI 3 into the downtown area. The city, as its name suggests, lies in far western Rhode Island along the west bank of the Pawcatuck River, opposite the Connecticut border. It's a pretty town with tree-shaded streets, elegant homes, and a long history. Besides sharing the placid river with Pawcatuck, its Connecticut twin, Westerly shares the same post office and railroad station. The town was supposedly founded by settlers John Babcock from Plymouth Colony and Mary Lawton from Newport. The lovers eloped from Newport in 1648. Other immigrants followed, settling on this riverbank purchased from local Indians. To tame the wild surroundings, a bounty of 20 shillings was given for every wolf killed in Westerly in 1687.

In later times Westerly flourished with a diverse industrial base, including shipbuilding, granite quarrying, and wool and cotton textile mills. The huge White Rock Company cotton mill, built in 1849 and purchased in 1874 for Fruit of the Loom, still testifies to this prosperity. Westerly's claims to fame now are the *Westerly Sun,* the only Sunday evening newspaper in the United States, and the world's oldest operating automatic telephone system. The city's downtown has a distinct small town flavor, dominated by 18-acre **Wilcox Park.** More than 100 tree and shrub species, an herb garden, flower beds, a large pond filled with goldfish, statuary including a marble Christopher Columbus, and a Braille-marked trail are found here. The **Wilcox Park Historic District** surrounds the park, with charming old buildings and stately Victorian residences. Buildings of note here are the 1872 Old Town Hall, the Romanesque-style Westerly Public Library, and the 1732 **Babcock-Smith House.** The house, open in summer with limited hours, is a well-preserved pre–Revolutionary War home furnished with period antiques that once belonged to Dr. Joshua Babcock, a member of the state's General Assembly, chief justice of the colony's Supreme Court, and a friend of Benjamin Franklin.

From Wilcox Park the drive begins by heading south on Beach Street (US 1A), following signs to Watch Hill. Solid old Victorian houses line the shaded street. The road passes River Bend Cemetery, with many monuments and gravestones of blue Westerly granite, and rolls south over worn granite hills covered with secluded homes. After almost 4 miles, bend right on Watch Hill Road. The asphalt twists southwest through hills for a few miles to **Watch Hill,** one of New England's oldest seaside resorts.

This small town at the state's far southwestern corner is surprisingly unpretentious and quiet. Parking is limited, however, and once the available spots are filled, it's hard to find anywhere to stop. Watch Hill received its unusual name during the Revolution when soldiers perched on this point and watched for British raiders. Later it became a fashionable resort. Many of its homes and estates with ornate hedges date from the 19th century. Bay Street, Watch Hill's business district, is lined with shops and restaurants. A statue of Ninigret, a 1630s chief of the Niantics, greets visitors on the street.

The **Flying Horse Carousel,** found at the end of Bay Street, is the oldest continuously operating merry-go-round carousel in the United States. The carousel, also called the Watch Hill

Carousel, was built in 1867, spent a dozen years traveling with a carnival, and was finally set up permanently in Watch Hill in 1879. The carousel, unlike modern ones, has no wooden deck. The 140-year-old, hand-carved, hand-painted horses are suspended from above and give the illusion of flying as the central shaft rotates. The horses were carved from single pieces of wood, decorated with agate eyes and genuine horsehair tails and manes, and have leather saddles. Adults are not permitted to ride the carousel; only children whose feet do not touch the ground when seated are allowed. The ride operates through the summer.

Another Watch Hill spot of interest is its popular sand beach facing the Atlantic Ocean. The bathhouse is near the carousel, and beach access is granted to those who have paid the bathhouse fee. A good short walk goes to the 1856 Watch Hill Lighthouse, an excellent example of a 19th-century lighthouse. The light is closed to visitors, but nearby is the **Watch Hill Lighthouse Museum.**

A longer hike is out to **Napatree Point Conservation Area and Beach.** The 1.5-mile walk to the very western end of Rhode Island follows a long spit of sand that juts into Little Narragansett Bay. It's a charming hike, with the emerald surf breaking on packed sand, a litter of shells sprinkled on the beach, and the wheeling of gulls on the sea breezes. Protected nesting sites of terns and ospreys are found on the point. Beach houses once lined this long, narrow stretch of land until a disastrous hurricane struck in 1938 and erased them. At the end of Napatree Point are the tumbled ruins of old Fort Mansfield, built during the Spanish-American War.

Watch Hill to Charlestown

To continue the drive, follow Watch Hill Road back to US 1A and go east (right) on Scenic Route 1A, also called the Shore Road Highway. Just down the road is the turnoff to 51-acre **Misquamicut State Beach** with a half-mile stretch of sand beach, and a couple of miles later the turn to Weekapaug. Down this lane is this small village to the east of Winnapaug Pond, a large lake and tidal marsh dotted with wading birds. Farther along the road is **Westerly Town Beach.** East of the town of Weekapaug is a good bluff viewpoint that overlooks the coastline and Block Island, a large offshore island that appears to float on the horizon.

Burlingame State Park, a 2,160-acre public area, sits on the left a few miles down the road. The park is a popular summer spot, with its 755-site campground and 600-acre Watchaug Pond. Boating and fishing are popular pursuits, and the Atlantic beaches are only a few miles to the south. The forested area also includes the Audubon Society's 29-acre **Kimball Wildlife Sanctuary,** a popular stopover for migrating birds and waterfowl including wood ducks, mallards, and teal. A nature center offers ongoing interpretative programs and displays.

Farther east, **Ninigret National Wildlife Refuge** and **Ninigret Park** border the highway on the south. Farther out along the coast is **Ninigret Conservation Area.** These three areas flank Ninigret Pond, a huge 1,700-acre salt pond separated from Block Island Sound by a long barrier beach. The wildlife refuge, on the site of an abandoned airport, offers rich habitat for migrating birds and waterfowl with deciduous forest, freshwater ponds, and salt grass. The refuge bird checklist includes 289 regular species (most of those found in the state) as well as 21 accidental species. The spring and fall migration seasons offer the best viewing, with the warbler migration peaking in May. Almost 3 miles of footpaths loop through the refuge. Ninigret Park, owned by the town of Charlestown, has swimming beaches, lifeguards in summer, a playground, bike course, and tennis and basketball courts. Ninigret Conservation Area fronts the ocean along the 2-mile stretch of barrier beach. Although it can be crowded, it is less busy than other nearby beaches because of parking limitations. When the area parking lot reaches its 100-car capacity, it is closed.

From the refuge the road becomes a four-lane, limited-access highway, and bends north to the site of **Fort Ninigret** on Fort

Neck Road. The bare outlines of this Dutch trading post from the early 1600s sit on the grassy north bank of sheltered Ninigret Pond. A large, irregular stone at the site is a memorial to the Niantic and Narragansett Indian tribes. The drive then enters **Charlestown.** Exit the highway for a quick stop in Charlestown center. The town, named for England's King Charles II, separated from Westerly in 1738. It grew as a mill village, grinding grains into flour and meal. An interesting stop here is the **Charlestown Historical Society Schoolhouse** on the grounds of the public library. The restored 1838 one-room school is austerely furnished with old desks, a potbelly woodstove, an old map, and blackboard. The school was used until 1918 in nearby Quonachantaug before being moved here. If you have time to explore the area, the local chamber of commerce has an information center that can point you in more directions.

A good detour drives south from town on Charlestown Beach Road. **Charleston Beach State Park** and **Charlestown Breachway** lie at road's end, where swimming and other water activities are the norm. The breachway also offers a 75-site campground for self-contained units.

Rhode Island's rich landscape was long inhabited by Native Americans, including the powerful Narragansetts. Today, the Narragansett Nation's tribal land covers more than 1,800 acres in the Charlestown area. The Narragansetts, whose place names are scattered across the region, have a historic village north of Charlestown off RI 2/112. Here is the reconstructed Greek Revival Narragansett Indian Church. Samuel Niles, the first native minister, was buried here in 1785. To the east, off Narrow Lane, is the Royal Indian Burial Ground, a 20-acre undeveloped site with the peaceful graves of Sachem Ninigret and his family from the 1700s. The **Narragansett Longhouse and Cultural Center** is the site of annual tribal celebrations. Nearby are the Cup and Saucer Rocks. These large boulders, balanced on a ledge, were rolled together to send rumbling messages.

Charlestown to Point Judith

Back on the drive, four-lane US 1 runs east from Charlestown through lush, deciduous woodlands broken by occasional pastures, farms, and houses. The next exit allows access to sandy Green Hill Beach and 800-acre **Trustom Pond National Wildlife Refuge.** The 160-acre pond here, the state's only undeveloped salt pond, is an important nesting and resting habitat for migratory birds. The refuge and pond, separated from Block Island Sound by sand barrier Moonstone Beach, offer diverse habitats for more than 300 identified birds and 40 mammals, including deer, red fox, mink, and otter. An information station is found on Matunuck Schoolhouse Road with directions for nature trails. The refuge is a serene oasis far removed from the nearby bustling shoreline and beaches. The Perryville State Trout Hatchery is just north of the highway.

Farther along the route is Matunuck Beach Road, which leads south to three popular beaches—Roy Carpenter's, South Kingston, and Mary Carpenter's—and the famous shore attraction, Theater by the Sea. Summer theater has been performed in this old barn, listed on the National Register of Historic Places, since 1933. **East Matunuck Beach,** a fine state beach, sits at the end of Succotash Road, accessed just up the drive. Nearby is Jerusalem and the Block Island Ferry Dock.

The highway arcs north around Point Judith Pond, a narrow bay, and passes **Wakefield.** This town and its neighbors to the north, Kingston and Peace Dale, are lovely, quiet villages set among verdant, rolling forests. Some excellent points of interest lie north of the drive here. Beginning in 1800, Wakefield thrived with the Narragansett and Wakefield mills. The village of **Peace Dale,** named by the village founder Rowland Hazard for his spouse, Mary Peace Hazard, is an old mill town on the Saugatucket River next to Wakefield. Hazard built several cotton and textile mills here that brought prosperity to Peace Dale through the 19th century. The mills contributed to the Civil War effort by weaving blankets for Union soldiers and by making khaki for World War I uniforms. The town houses the Museum of Primitive Art and Culture with its small but fascinating collection of Rhode Island Indian artifacts, as well as items from other Native American tribes and world primitive cultures.

A few miles north of Peace Dale is **Kingston.** The town, originally part of a land tract purchased from the Narragansetts in 1658, was named Little Rest until 1825. Kingston has a charming historic district with numerous old buildings and homes, including the 1820 Kingston Congregational Church; the 1710 John Moore House, the oldest here; the 1820 George Fayerweather House, built by an Afro-Indian blacksmith who was the son of a slave; the 1792 Washington County Jail; and the landmark Kingston Railroad Station. The University of Rhode Island's campus is also in Kingston.

Just west of Kingston is the 3,349-acre **Great Swamp Management Area,** the site of the Great Swamp Fight. This bloody, decisive battle fought on December 19, 1675, was a turning point in King Philip's War. The Wampanoag Indians and their chief, King Philip, tried to force settlers off Indian lands. Colonists marched on a Wampanoag Indian encampment set on a frozen island in the Great Swamp and set fire to an Indian fort, killing more than 500 people including warriors, women, and children. King Philip, however, was not at the camp and continued to fight until his capture and death in August 1676. A hard-to-find historical marker in the woods commemorates the battle.

The drive route exits from US 1 just past Wakefield and turns south (right) on RI 108, the Old Point Judith Road. The drive follows RI 108 past houses, cottages, and businesses for 4 miles to US 1A (Ocean Road). Cove Road, just before the highway intersection, goes west to the active fishing village of Galilee, some good seafood restaurants, and Salty Brine and Roger W. Wheeler Beaches. Another side road makes a jaunt over to **Fisherman's Memorial State Park,** a popular camping area.

From the junction of RI 108 and US 1A, a right turn leads drivers a short distance to **Point Judith** and its lighthouse poised on the land's jutting tip. The lighthouse and adjoining Coast Guard station are surrounded by a low wall and a beach of broken boulders. The 51-foot-high, octagonal lighthouse, built in 1816, sits on a strategic site overlooking the entrance to Narragansett Bay. Not only was this point an important lookout in the Revolution, but its lighthouse beacon has warned sailors of impending peril since first placed here in 1810. So many ships have wrecked off Point Judith that it acquired the nickname "Cape Hatteras of New England." Spectacular views of the ocean wrapping around this windswept, rocky point are seen from a nearby hillock. This is a great spot to watch the endless waves breaking on the shoreline. Fishing boats dot the undulating water; the white cliffs of Block Island puncture the southwestern horizon; and large container ships sail slowly toward New York.

Along Narragansett Bay

From Point Judith the drive runs north on US 1A for 6 miles to Narragansett Pier. Cottages and beachfront homes line the highway. Several great sand beaches lie along Narragansett Bay and the road. The best is **Scarborough State Beach,** a 300-yard stretch of beach and dunes that is very popular in summer with swimmers and tanners. The 3,000-car parking area is usually full on weekends. On the southern outskirts of Narragansett Pier is the Ocean Road Historic District.

US 1A wends slowly through the resort town of **Narragansett Pier.** The town was a fashionable resort in the late 19th century, with luxurious hotels and the famed Narragansett Casino. This ornate casino was built in 1883, but was destroyed by fire in 1900. A stopover for the well-heeled, traveling between Newport and New York, the grand casino offered bowling alleys, billiard halls, tennis courts, a rifle range, theater, and ballroom. All that remains of it today is The Towers, a local landmark spanning

Ocean Road. The local chamber of commerce is housed in the base of the twin stone turrets. On the north side of town are a couple of beaches, including crescent-shaped **Narragansett Town Beach,** and the **Narragansett Indian Monument.** This mammoth 23-foot-high sculpture weighs 10,000 pounds and is carved from the trunk of a huge Douglas fir. **South County Museum,** sitting on US 1A north, details the area history with artifacts, tools, and exhibits including a country kitchen, cobbler's shop, tack shop, children's nursery, general store, and antique vehicles such as a horse-drawn hearse and fire engines.

US 1A, here called Boston Neck Road, runs north and crosses the broad Pettaquamscutt River before entering a lovely stretch of countryside. Farms and old houses, stone walls, swampy lowlands, and dense forests cover the land. This area along the west coast of Narragansett Bay was the site of many of the large Rhode Island plantations that flourished in the 1600s and 1700s. The **Silas Casey Farm,** on the left, was once a plantation and is still a working 350-acre organic farm surrounded by stone walls. The house, built in the mid-1700s, is furnished in period pieces from the Casey family. During the Revolutionary War, an encounter between colonial patriots and British sailors led to a musket shot. The hole still pierces the parlor door.

The **Gilbert Stuart Birthplace and Museum** is just north of the farm and a mile west of the highway on Snuff Mill Road. Stuart, America's most famous 18th-century portrait painter, was born here in 1755 in a red Colonial house above his father's snuff mill. After showing an early aptitude for art, Stuart went to England and studied. He returned to America in 1792 and painted famous 18th-century Americans, including Presidents John Adams, John Quincy Adams, Thomas Jefferson, James Madison, and, most notably, George Washington. The best known of Stuart's 124 portraits of the first president is the unfinished *Athenaeum Head,* which appears on the face of the one-dollar bill. After painting more than a thousand portraits, Stuart died in 1828.

The drive continues north and in less than a mile reaches RI 138. Turn right on RI 138 and drive over the sweeping 1.5-mile Jamestown Bridge to **Conanicut Island.** The highway dashes across the island, but a loop drive around the island on East Shore, North, and Beavertail Roads makes a good interlude. The roads thread through a bucolic countryside, passing old farms, lush forests, and green pastures. Several interesting stops are found along the way.

Conanicut Island

The island, named for the Narragansett Sachem Canonicus, stretches almost 9 miles from Conanicut Point on the north to Beaver Tail on the south. Conanicut, acquired from the Indians in 1657, sits at a strategic location at the entrance to Narragansett Bay. Quaker farmers settled the island, tilling its fertile soil and raising sheep. During the American Revolution, the British captured and occupied the island in 1776 until 4,000 French soldiers overran it in 1778 and forced the British and Hessian troops to retreat to Newport.

Conanicut Island points of interest include the Watson Farm, Sydney L. Wright Museum, Fort Wetherill State Park, and Beavertail Lighthouse and State Park. The **Watson Farm,** now owned and operated by the Society for the Preservation of New England Antiquities, was a family farm founded in 1798. Five generations of Watsons lived here over 183 years. Today the farm is run as a living history site. Nearby is the old Jamestown Windmill. The **Wright Museum,** tucked into a corner of the Jamestown library, is an interesting collection of Native American artifacts. Many were recovered from the West Ferry archaeological site.

Fort Wetherill, perched atop a rocky crag opposite Newport, has long protected this narrow strait called East Passage. The first fort was installed in 1776. Later forts were equipped with powerful guns to protect the harbor in the Civil War and both world wars. Nearby Pirate's Cove is the possible burying place of a treasure belonging to the notorious buccaneer Captain Kidd, who often visited fellow pirate Captain Thomas Paine in Jamestown. Fort Getty sits on the opposite site of the island. Beavertail State Park, encompassing the island's rocky tip, offers what is perhaps Rhode Island's best ocean view. The surf endlessly pounds against the cliffs and boulders along the shoreline here, and sunsets are usually spectacular. The nearby Beavertail Lighthouse has guided mariners through these treacherous channels since 1749. A small museum details the colorful history of Rhode Island lighthouses.

Newport

The last drive segment begins on the east side of the island. Get back onto RI 138 east and follow the traffic over the **Newport Bridge,** a towering suspension bridge arcing over the East Passage. The view from this lofty span is stunning, with the whole bay and its islands spreading out below. After a couple of miles, the bridge gently descends and enters the famed resort and sailing city of Newport.

Spread across a J-shaped promontory at the southern end of Aquidneck Island, **Newport** is one of America's most beautiful and historic cities. History lurks at every turn. Rather than being encased in a museum, the town wears its past like an opulent cloak. It was settled in 1639 by religious dissenters from Massachusetts. The city acquired a reputation as a haven for tolerance, welcoming Quakers, Jews, and religious outcasts from the more rigid northern colonies.

Newport, with its ocean-edge location and good, deep-water harbor, flourished as a trade center and major port that shipped goods and livestock from the American colonies to Europe, including horses, wool, agricultural products, and rum. It was rum that put this bastion of tolerance at one corner of the infamous Triangle Trade between Newport, West Africa, and the West Indies. Newport rum was traded for black slaves in Africa, who in turn were traded in the Caribbean for sugar and molasses to make more

Stately 19th-century summer "cottages" line the Cliff Walk in Newport, including Ochre Court, the second-largest mansion in Newport.

rum in Newport. As many as 60 ships based out of Newport participated in the slave trade. Newport was also a popular port for privateers, ships that were licensed by the state to destroy competing enemy ships. The rum trade established the pineapple as the Newport symbol for hospitality. A returning seaman, when he was ready to receive visitors, would place a pineapple outside his door.

Newport's strategic location and trade also made it an important site in the Revolution. The British captured and occupied the city from 1776 to late 1779, when the colony's French allies liberated it. The occupation, however, decimated the city's population and destroyed many of its buildings. It wasn't until around 1830 that Newport began recovering from its war wounds when Southerners began summering here. By the late 19th century, Newport was recognized as "America's First Resort" and became the summer playground for the rich and famous. During this gilded age the country's robber barons and wealthy industrialists, with immense profits and no income

tax, bought the city's beaches and cliff-top properties as sites for their "cottages." These cottages, numbering almost 60, were huge, extravagant mansions and estates that are reminiscent of the chateaux and castles of European royalty. Some cost as much as $10 million to build and furnish. Their owners, who came only for, at most, 11 weeks of the year, spent hundreds of thousands of dollars entertaining during the summer season, and thought that because they had money they were the American aristocracy. One society matron went so far as to call the local townspeople "footstools." Taxes and the Great Depression brought this lavish age to a grinding halt, and the summer cottages became too expensive for even the very rich to maintain. By 1950 only a few remained in private hands.

Today Newport, one of the world's yachting capitals, is a charming and attractive city with lots of historic attractions, festivals, and events. The drive winds through the city's downtown. After crossing the bridge, head south on RI 238 (Farewell Street). Farther along, turn left on Memorial Boulevard, which becomes RI 138A. Drive eastward on the tree-lined street and park on the east side of Newport along a breakwater that separates Easton Pond on the left from the ocean waters. The famed Cliff Walk footpath begins on the south side of the road.

The **Cliff Walk** is a 3-mile trail that winds along the cliff top between the breaking waves of Rhode Island Sound and a long row of seaside mansions. Six of the best mansions, managed by the Preservation Society of Newport County, are open for public viewing. The society began acquiring these mansions, including The Breakers, Rosecliff, and The Elms, in 1945. These huge palaces are Newport's single most popular attraction, visited by more than a million people annually. The Cliff Walk and a visit to at least a couple of the mansions is a requisite for any Newport visitor.

The Cliff Walk at Newport offers spectacular views of waves breaking on the rocky shore.

The trail winds above the broken cliffs, offering marvelous views of the surf breaking on the rocks below. The mansions, facing the ocean, line the path. All the mansions open for tours are on Bellevue Avenue, paralleling the Cliff Walk to the west. One of the best and the largest is **The Breakers,** the 1893 summer home of railroad magnate Cornelius Vanderbilt. The 4-story limestone mansion, surrounded by an 11-acre manicured estate, contains 70 rooms. It took hundreds of craftsmen 2 years to build, with entire rooms built in Europe and shipped over to Rhode Island.

To continue the drive, head south from Memorial Boulevard on Bellevue Avenue past the mansions to Ocean Avenue and the Ocean Drive along the south coast of the island. This route yields inspiring scenic vistas of Rhode Island Sound and the rocky shoreline, passing Brenton Point State Park, Hammersmith Farm, and Fort Adams State Park. The drive follows Ridge Road, Harrison Avenue, Wellington Avenue, and Thames Street back to Memorial Boulevard and the drive's end in downtown Newport.

Lots of other activities and attractions are available for Newport visitors. Museums include the International Tennis Hall of Fame, Museum of Newport History, Newport Art Museum, Newport Historical Society, and the Museum of Yachting. There is Colonial Newport; the old Common Burial Ground; the 1687 White House Tavern, thought to be the oldest tavern in the United States; and many stores, shops, and restaurants. Popular events include the annual Newport Music Festival in July, the Newport Folk Festival in August, and the Classic Yacht Regatta over Labor Day weekend. Newport, the City by the Sea, makes a fitting climax and busy ending to Rhode Island's magnificent coastal drive.

6 Western Rhode Island

General description: A 31-mile open loop drive across rolling, forested hills in western Rhode Island.

Special attractions: Arcadia Management Area, Tomaquag Museum of the American Indian, Parker Woodland Wildlife Refuge, Swamp Meadow Covered Bridge, Scituate Reservoir, Clayville National Historic District, Prinster-Hogg Park, hiking, birding, boating, fishing.

Location: Western Rhode Island between I-95 and Slatersville.

Drive route number and name: RI 102 (Victory Highway).

Travel season: Year-round.

Camping: No campgrounds along the drive. George Washington Management Area west of Chepachet has a 45-site camping area.

Services: Limited services on the route at Clayville and Chepachet. Full services, including gas, food, and lodging, in Slatersville at the northern terminus.

Nearby attractions: Blackstone Gorge State Park, Providence attractions, Snake Den State Park, Lincoln Woods State Park, Diamond Hill State Park, Slater Mill Historic Site, Narragansett Bay area, Newport, Cliff Walk, Great Swamp Fight Monument.

The Route

The 31-mile Western Rhode Island Scenic Route, following RI 102, slices across the west side of the state through a rural, hilly region far removed from the glamour and glitz of Newport. Rhode Island, smallest of the 50 states, covers a scant 1,210 square miles, an area only slightly larger than half the size of Delaware, the next smallest state. The state, however, is densely populated. Its 1,018 people per square mile are exceeded only by New Jersey, the most densely populated state. Nine-tenths of the Rhode Island population lives in urban areas such as Providence, New England's second-largest city. This drive crosses a more sparsely populated region of low rolling

Western Rhode Island

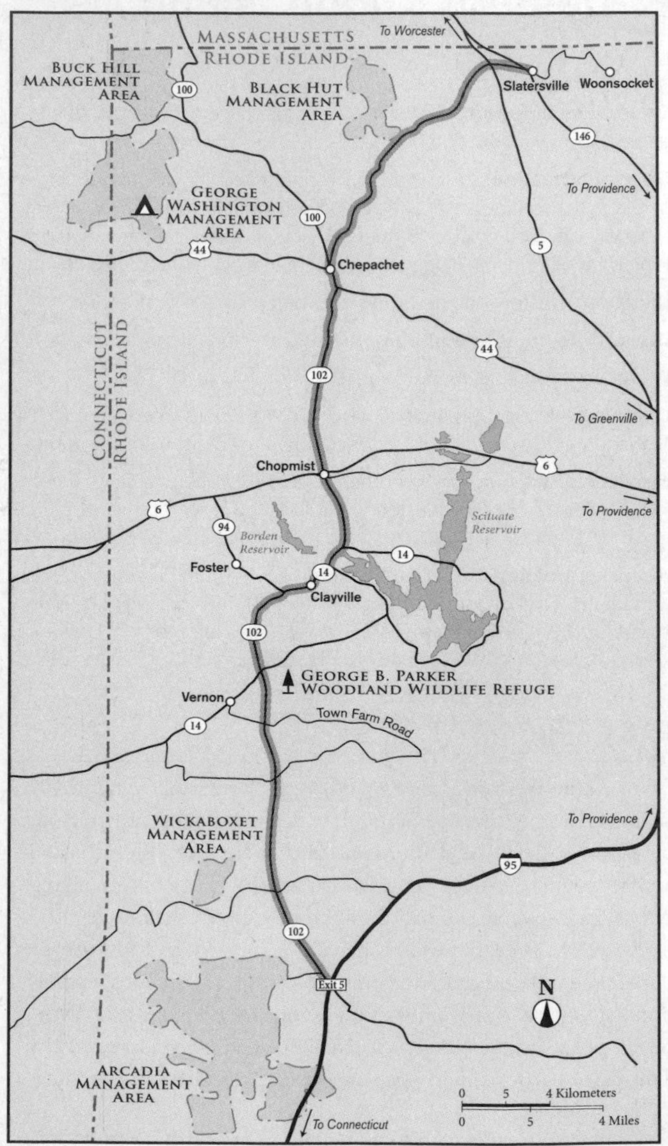

hills, part of what's called the Eastern New England Upland by geographers. The state's high point, 812-foot Jerimoth Hill, rises west of the drive near the Connecticut border.

I-95 to Clayville

Easily accessed from Providence, the drive route begins at exit 5 on I-95. Turn onto RI 102 north. The two-lane road runs north through rolling hills thickly matted with white pine and hardwoods, including oak, hickory, beech, and maple. Occasional homes break the forest solitude. Southwest of the highway sprawls **Arcadia Management Area,** with over 14,000 acres, it's the state's largest reserve of public lands. The irregular-shaped unit, managed by the Rhode Island Department of Environmental Management, offers numerous outdoor opportunities along its more than 30 miles of maintained trails, including canoeing and swimming on Browning Mill Pond Recreation Area, birding, hunting, and excellent trout fishing in Wood River. Two good hikes are the 6-mile Arcadia Trail and the 3-mile Mount Tom Trail along the river.

Near the town of Arcadia is the interesting **Tomaquag Indian Memorial Museum.** The museum, run by the Narragansett tribe, offers an intriguing collection of artifacts, tools, and photographs that primarily detail the prehistory and history of New England's diverse native cultures. Another good side trip is to **Step Stone Falls,** a picturesque but obscure Rhode Island natural wonder. Here the Fall River cascades over a series of broken rock ledges. The falls are north of Escoheag on rough Falls River Road.

The main drive route runs past the University of Rhode Island's Alton Jones Campus. About 3 miles west of the highway off Plain Meeting House Road is 678-acre **Wickaboxet Management Area.** Low, rolling hills covered with hardwood trees and dense brush, with a trickling brook, lie in this small, rarely visited preserve. This woodland area is good for birding and is popular with hunters in autumn.

The highway continues north and a few miles later reaches **George B. Parker Woodland,** an 860-acre Audubon Society nature area. The preserve offers what a kiosk sign calls "historic archaeology." The area, now blanketed with second-growth hardwoods, was a farming community called Coventry Center in the 1700s. The land was originally purchased from the Narragansett tribe in 1642 and passed through a succession of owners until George Parker acquired it and deeded it to the Audubon Society in 1938. More than 7 miles of trails lace the parkland, allowing excellent opportunities to study the area's flora and fauna or simply soak up the peace and quiet. The area protects an almost pure 15-acre stand of chestnut oak.

Back on the drive, the road runs through woods interrupted by overgrown stone walls that once marked a farm's cleared pastures. Five miles from the wildlife refuge, the highway enters **Clayville,** a national historic district of old houses. Foster lies a couple of miles west of Clayville on RI 94. Just north of Foster off Central Pike is Rhode Island's only covered bridge. The bridge, a replica of a classic 19th-century covered bridge design from 1820, was first built over Hemlock Creek in 1992. Vandals torched it the following year. The town pulled together, solicited donations of lumber, money, and time, and rebuilt the **Swamp Meadow Covered Bridge** in 1994.

Clayville to Slatersville

The highway twists through Clayville and drops northeast to the Ponaganset River branch of massive **Scituate Reservoir.** The Y-shaped reservoir, built in 1915 for Providence's drinking water supply, covers 13,000 acres in the Pawtuxet River drainage and is the state's largest freshwater lake. The lake here is a narrow arm, fringed with tall trees. After crossing a bridge over the water, RI 102 and 14 intersect. Keep left on RI 102. **Prinster-Hogg Park,** with picnic tables and grills, sits at this lonely intersection. The park is named for two pilots who landed a burning

Dense woodlands line the glassy waters of Scituate Reservoir, Rhode Island's largest lake.

airplane on the reservoir in 1982, saving themselves and 12 passengers.

The road heads over rolling hills blanketed with woodlands and occasional farms with open fields for the next 6 miles to the town of **Chepachet.** This charming town, the largest along the drive, sits at a crossroads of three highways. It's a pleasant, rural town with antiques shops and historic homes and buildings lining its main street. The **Job Armstrong Store** houses a small museum, run by the Gloucester Heritage Society, which displays local artifacts and lore every Saturday. Of particular interest are the details of the Dorr Rebellion, a suffrage revolt led in 1842 by local resident Thomas Dorr.

Dorr had been elected governor of Rhode Island, but the incumbent governor Samuel King refused to concede the reins of government. The state militia entered the fray and put down the rebellion. The failed revolt, however, led to a new state constitution and liberalized voting rights in 1843. Rhode Island at that time was still governed by its original 1663 charter, which allowed voting only by landholders or their eldest sons. The balance of power in the state government had been slanted toward rural areas, whereas most of the population lived in cities and was denied voting rights. The Dorr Rebellion helped remedy that situation by allowing the vote to native-born men who paid $1.00 or more in taxes annually or served in the state militia.

Jerimoth Hill, Rhode Island's 812-foot high point, is 500 feet south of RI 101 about 10 miles west of the drive. To climb the high point, head west on RI 101 to the crest of hill with a sign that designates Jerimoth Hill. The actual high point, a humped boulder, is reached by a short trail that heads south from the highway.

The drive continues north from Chepachet on RI 102 for its last 10 miles. It passes an old cemetery and Sucker Pond before arcing northeast through thick forests in the hills west of the Chepachet River. A few large natural areas lie in the northwestern corner of the state, west of the highway. **Buck Hill Management Area** is an undeveloped 2,049-acre tract of dense forest in the state's extreme corner. Reached via RI 100 and Buck Hill Road, it offers some fine hiking and lots of wildlife and waterfowl. **George Washington Management Area**'s 4,000 acres lie west of Chepachet. This hilly parkland, with some points reaching as high as 770 feet, is broken by rocky outcrops, large **Bowdish Reservoir** with a beach and boating, and a 55-site campground. The area also has hiking trails, swimming, ski touring, fishing, and hunting in season. **Black Hut Management Area** sits just north of the drive and Glendale. The 1,548-acre area, managed by the state Division of Fish and Wildlife, is a quiet woodland reserve spread over low hills. More than 5 miles of trails thread through the forest. Nearby is Spring Lake, a popular fishing and boating pond.

Houses and businesses abut the pavement as the drive nears Slatersville and its end at four-lane RI 146. **Slatersville,** along with other northern Rhode Island towns situated along rivers, was

A canopy of trees arches over a country lane in northwestern Rhode Island.

a birthplace of America's Industrial Revolution, the far-ranging economic transformation that changed the country from a nation with a rural, agrarian base to an industrial, urban power.

In 1793 Samuel Slater opened the Slater Mill along the Blackstone River at Pawtucket southeast of here. This was the first modern textile mill in the United States to be powered by water. An English textile machinery expert, Slater had departed England disguised as a farm laborer. The English, after inventing power spinning and weaving, had restricted emigration of textile workers to keep trade secrets from rival countries. Slater, however, reinvented the machines from memory after arriving in America, setting the stage for the complete transformation of the US textile industry from a home craft to mechanized, mass production.

The town of Slatersville, a pleasant old town founded in 1807, was a planned community built around a mill that was overseen by John Slater, Samuel's brother. The Slaters erected houses, churches, and schools for the mill workers. Points of interest include the triangular village green with the circa 1838 Greek Revival–style Congregational Church alongside, the **William Slater Mansion,** and the **Slater Mill.**

Northeast of town is the hard-to-find **Blackstone River State Park.** The Blackstone River has excavated an impressive little gorge through the bedrock here, offering scenic views and the state's only whitewater canoeing. Ask at Slatersville for precise directions. Otherwise, the drive dead-ends against RI 146, a four-lane, divided highway that quickly leads south to Providence and its numerous attractions.

APPENDIX: SOURCES OF MORE INFORMATION

For more information on lands and events, please contact the following agencies and organizations.

Connecticut

Connecticut Commission on Culture and Tourism
1 Constitution Plaza, 2nd Floor
Hartford, CT 06103
(888) 288-4748 (CTVISIT)
ctvisit.com

Connecticut State Parks Division Bureau of Outdoor Recreation
79 Elm St.
Hartford, CT 06106
(860) 424-3200, (866) 287-2757
dep.state.ct.us/stateparks/

Greater New Milford Chamber of Commerce
11 Railroad St.
New Milford, CT 06776
(860) 354-6080
newmilford-chamber.com

Mystic Seaport
PO Box 6000
75 Greenmanville Ave.
Mystic, CT 06355
(860) 572-5302, (888) 973-2767
mysticseaport.org

Northeastern Connecticut Chamber of Commerce
3 Central St.
Danielson, CT 06239
(860) 774-8001
nectchamber.com/

Old Saybrook Chamber of Commerce
PO Box 625
146 Main St.
Old Saybrook, CT 06475
(860) 388-3266
oldsaybrookchamber.com

Western CT Visitors Bureau
PO Box 968
Litchfield, CT 06759
(860) 567-4506
litchfieldhills.com

Rhode Island

Blackstone Valley Tourism Council
175 Main St.
Pawtucket, RI 02860
(401) 724-2200, (800) 454-BVTC
tourblackstone.com

Charlestown Chamber of Commerce
4945 Old Post Rd.
Charlestown, RI 02813
(401) 364-3878
charlestownrichamber.com

Narragansett Chamber of Commerce
PO Box 742
36 Ocean Rd.
Narragansett, RI 02882
(401) 783-7121
narragansettcoc.com

Rhode Island Tourism Division
315 Iron Horse Way, Ste. 101
Providence, RI 02908
(800) 250-7384
visitrhodeisland.com

INDEX

THOR

Stewart con-
tract ... t. He's
writ... ud-
ing S... o-
shire ... K
Rock
Rock ... w
Eng ... d
Best ... onal
clim... do
and
.wix
pho